# symposia on theoretical physics and mathematics

**8**

*Contributors to this volume:*

T. Bressani
G. Charpak
N. Dallaporta
S. G. Deo
C. J. Eliezer
J. Favier
A. S. Galiullin
J. I. Horváth
C. Joachain
L. Massonnet
P. M. Mathews
A. Mercier
W. E. Meyerhof
K. S. Padmanabhan
N. Prakash
I. V. V. Raghavacharayulu
A. Ramakrishnan
M. Rho
H. S. Shapiro
N. V. Subrahmanyam
J. H. Williamson
Č. Zupančič

# symposia on theoretical physics and mathematics

Lectures presented at the
1967 Fifth Anniversary Symposium
of the Institute
of Mathematical Sciences
Madras, India

Edited by
**ALLADI RAMAKRISHNAN**
Director of the Institute

PLENUM PRESS • NEW YORK • 1968

ISBN 978-1-4684-7723-8      ISBN 978-1-4684-7721-4   (eBook)
DOI 10.1007/ 978-1-4684-7721-4
Library of Congress Catalog Card Number 65-21184


© 1968 Plenum Press
Softcover reprint of the hardcover 1st edition 1968
A Division of Plenum Publishing Corporation
227 West 17 Street, New York, N. Y. 10011

# Introduction

This volume comprises the lectures given at the Fifth Anniversary Symposium held at the Institute of Mathematical Sciences, Madras, India, during January 1967.

Professor Dallaporta of Padua delivered the inaugural address on the fundamental problem of quasars "whose study appears to hold implications for cosmology." He presented a critical review of several attempts to understand their exceptionally large red shifts and also discussed the physical theories concerning the cause of the explosions which give rise to the quasars and to the tremendous energy output they require – questions which still remain unanswered. He stated, in concluding, that we may have to invoke certain aspects of the present theories of elementary particles in order to unravel these mysteries.

Professor Mercier, well known for his studies on the philosophical foundations of modern physics, critically examined the various attempts, such as that of Einstein, to formulate a unified field theory. Ramakrishnan and Raghavacharayulu described the group theoretical significance of the hierarchy of $L$-matrices and the $\sigma$-operation introduced earlier by Ramakrishnan, and Mathews spoke about quantizing the relativistic "Schrödinger type" equations for higher spin which were described in Volume VI of these proceedings. Horváth lectured on his contributions to a relativistic generalization of the kinetic theory of gases, in which the relativistic phase-space represents a geometrization of the dynamics of gases. Jochain's paper was devoted to a comprehensive review of both the variational and minimum principles in scattering theory and their utility in obtaining bounds on scattering length and phase shifts. Charpak presented the results of recent experiments at CERN on the absorption of pions by a pair of nucleons and discussed its significance for the study of nuclear structure. Rho analyzed muon capture as a means of probe nuclear structure.

In his lectures on Raikov systems, Williamson described how they "provide yet another instance, at a quite elementary level, of just how complicated a system the real numbers are." With their origin in a study by D. A. Raikov of the structure of the measure algebra $\mathfrak{M}(R)$ of real line $R$, the main interest of the subject is perhaps due to its bearing on the structure of $\mathfrak{M}(R)$ or, more generally, of $\mathfrak{M}(G)$, where $G$ is a locally compact Abelian topological group. Williamson also stressed the need for studying Raikov systems for their own sake, independent of any possible applications. The ideas and results described here are still in a state of development.

Shapiro proposed a formal definition of generalized analytic continuation and gave various examples of function classes which admit generalized analytic commutation in this sense. Later he discussed such topics as continuations by matching boundary values and the Laplace transform of almost periodic functions.

The other contributions on mathematical topics were by Eliezer on convex functions and related inequalities, Deo on functional differential equations, Nirmala Prakash on harmonic differential forms in a general manifold, Padmanabhan on developments in the theory of univalent functions, and Subrahmanyam on congruent embedding in Boolean vector spaces. The volume also includes a contribution on the construction of total programmed-motion systems by Galiullin.*

*Alladi Ramakrishnan*

---

*The original manuscript in Russian was translated by the Publishers, Plenum Press, N. Y.

# Contents

# Contents of Other Volumes

# VOLUME 3

# VOLUME 4

# VOLUME 5

# VOLUME 6

# A General Outlook on the Quasars

N. DALLAPORTA

*UNIVERSITY OF PADUA*
*Padua, Italy*

---

Since my main field of research is astrophysics, I suppose that when asking me to present this paper, Professor Ramakrishnan probably intended to stress the point that the Institute of Mathematical Sciences, which is so rapidly and vigorously developing from year to year, after having extended the domain of its activities from theoretical physics to pure and applied mathematics, is now just starting to include astrophysics. And therefore, the real meaning of my paper should be to show that this new extension is worthwhile and that astrophysics may actually provide the theoretical physicist with problems which are no less important, exciting, and fundamental as those related to elementary particles. For this reason, I have chosen as the subject of my paper the problem of the so-called quasi-stellar sources, or quasars, whose study appears to hold implications for cosmology and the type of model which we use to explain our world. Therefore, I shall try, in a summary of this field, to present its most outstanding features, and the important steps in our knowledge of the universe which have been accomplished by the radioastronomical techniques developed in the last decades, among which the discovery of the quasars appears actually to be one of the most significant.

As is well known, powerful radiotelescopes have permitted us to demonstrate the presence of sources in the sky emitting on the range of the cm wavelength, and measurements on several wavelengths have yielded information on their spectra. In most cases

1

these spectra are not thermal, but follow a power law of the type $S_\nu \sim \nu^{-\alpha}$, typical of the so-called synchrotron phenomenon, that is, deceleration of relativistic electrons moving in magnetic fields.

The question then was where are these fields and electrons generated?

An answer to it has been suggested by the galactic radio sources, especially the Crab nebula. This strong source is located in the exact position where a supernova exploded in the year 1054. Thus, we know that a radio source is what is left from a star's explosion about a thousand years after the outburst. This result, obtained on a stellar scale, has been extended and extrapolated to interpret also the extragalactic radio sources, most of which, as observation reveals, are connected to the central parts, or nuclei, of galaxies. We know today that many galaxies are radio sources, including ours; however, most of them are only weak sousces (output between $10^{38}$–$10^{41}$ ergs/sec) and only few, mostly giant elliptical galaxies, are strong ones (output between $10^{41}$–$10^{44}$ ergs/sec). According to the previous extrapolation, radio sources in galaxies are now thought to be the remnants of explosions which have occurred in their nuclei, and a whole scale of different galaxy types are now interpreted as showing the different phases of such explosions:

*Initial phase:* A bright and concentrated nucleus, with little radio emission (Seyfert galaxies).

*Central phase:* Evidence of output of gases and material from the nucleus with increased radio emission (M 82).

*Final phase:* Characterized frequently by two radio sources situated generally on the normal to the plane of the galaxy; as the distance of the sources increases, the output of the radio power declines.

Moreover, there are several reasons to suppose that these explosions are probably recurrent, with a period of the order of $10^6$–$10^7$ years. It may also be remarked that they appear to occur only in the largest and brightest of the galaxies.

It is in such a general frame that the discovery of the quasi-stellar sources (QSS) has to be located.

As soon as very precise measurements concerning the exact position of radio sources became available, it was discovered by Sandage that some sources (3C 48, 196, 286), which had escaped identification with any known galaxy, were practically coincident with point source images completely similar to stars. Hence, the name of QSS or quasars given to these sources. More cases were later gradually found, so that on the whole about 80 of such sources have now been recognized. Although they may differ in several details from each other, one may try to summarize their general characteristics as follows:

*Radio data:* For about half of the sources (57%) the radio spectrum follows the power law of the synchrotron type $\nu^{-\alpha}$. For about 25% the spectrum has an upward convex curvature, which is typical of the self-absorption phenomenon of the synchrotron radiation by the electrons themselves. The frequency corresponding to the maximum of the spectrum $\nu_m$ depends on the magnetic field $H$ and the angular diameter of the source $\theta$ as $\nu_m \sim (H^{1/2} \theta^{-2})^{2/5}$. For a remaining 10%, the curvature is opposite (convex downward), and this corresponds to a more complex case to be mentioned later. The remaining 8% are still unclassified.

A further remark of interest is that the radio source is not always centered on the optical point source. There are several cases of double radio sources, more or less symmetrically disposed with respect to the optical image, as for galaxies. Moreover, one sometimes sees jets or wisps emanating from the point source and connected to the radio spectrum.

*Optical data:* The spectrum consists of a continuous background with broad emission lines. Only recently, Oke has succeeded in separating the contributions of these two spectra for some of the most luminous sources. It has thus turned out that the continuous spectrum may also be fitted by a kind of power law, either straight or with some curvature, and this again stresses that, as for radio spectrum, these spectra are not thermal but are compatible with being either synchrotron radiation or emission of quanta in free–free collisions. In some but not in all cases, it has been possible to fit both the optical and the radio spectra with a single power law. For what concerns the color, the main characteristic is an ultraviolet excess in respect to normal main sequence stars, to be found only in some peculiar stellar states as white dwarfs or old novae.

The emission line spectra have for a long time been a puzzle, as no one was able to identify these lines with those of any of the known elements. Finally, Schmidt succeeded in recognizing the Balmer series and several forbidden lines of different origin on the spectrum of 3C 273, with the assumption of a high red shift

$z = \Delta\lambda/\lambda$ of 0.158; and this was the key to the general understanding of the phenomenon. However, it turned out that the red shifts, required for the interpretation, were on the whole extremely high (up to $z \simeq 2$), even with respect to the highest red shifts related to the expansion of the universe observed for the most remote galaxies ($z \sim 0.5$).

Thus, it became evident that the outstanding feature of the quasars were their exceptionally large red shifts, and all the interest concentrated in trying to understand their origin.

A first assumption was that the red shift could be gravitational, a well-known consequence of general relativity being that the spectral lines of a source of mass $M$ and radius $R$ are shifted toward the red by an amount of

$$z = \frac{\Delta\lambda}{\lambda} = \frac{G\,M^2}{c^2\,R}$$

(where $G$ is the gravitation constant and $c$ is the velocity of light), which may turn out to be very large if a high mass is concentrated in a small volume. In an important paper related to the two best observed cases, 3C 48 and 3C 273, Greenstein and Schmidt were able to show, by an argument which admits of no brief summary, that it was extremely unlikely, when assuming the quasars to be either dense stars or compact galaxies, that the gravitational effect could explain the red shift, and their conclusion was generally accepted.

The red shift had then to be thought as due to the Doppler effect, and the simplest assumption (as all observed shifts were red and none blue) was to suppose it was cosmological as for galaxies. It turned out on this line that the distances of 3C 48 and 3C 273 were of the order of 1000 Mpc and their absolute luminosity of $\sim 10^{46}$ ergs/sec, that is, 100 times greater than for the most luminous galaxies and, therefore, by far the most luminous objects in the world.

The further consequences of the cosmological assumption were discussed by Greenstein and Schmidt,[1] and may be summarized as follows: The radius of the radio emitting region can be deduced to be of the order of 1000 pc (small in comparison with the radius of 15,000 pc of our galaxy). The seat however of the optical lines is much narrower and of the order of only 1–10 pc. Further, it is likely that the continuum is emitted by a still narrower region. The quasar thus appears as an exceedingly concentrated structure (on galactic scale) formed by different shells in different physical conditions. If one assumes further that the widths of the emission lines are due to expansion, and from this resulting expansion velocity and the value deduced for the diameter of the quasar one calculates the expansion time, one then obtains an age of only $10^3$ years. However, the time necessary for secondary structures of the quasar, such as the jet and the wisps, observed for the two objects under study, to get where now they are if supposed to have been ejected from the central structure, amounts at least to $10^6$ years. If then one assumes this to be the minimum lifetime, the total amount of energy irradiated up to now turns out to be of the order of at least $10^{60}$ ergs, and in order that such a quantity may be provided by conventional sources, such as nuclear reactions or gravitational potential, one needs to assume for the quasar a mass of at least $10^9$ solar masses, that is, a value on the galactic scale (in comparison our galaxy has a mass of $10^{11}$ solar masses).

The physical picture just arrived at under the cosmological assumption suggests several similarities of quasars with galaxies, namely the following:

1. The same type of emission lines as the nuclei of Seyfert galaxies.

2. The emission of jets, found also in some galaxies (M 87).

3. Similarity in the structure of the double sources.

4. Similar intensity of radio emission as for strong radio galaxies.

5. Similarity with the nucleus of the so-called $N$ or compact galaxies.

However, besides these analogies, one also has to stress some major differences, which are:

1. The optical luminosity is higher by a factor of 100 with respect to the most luminous radio galaxies.

2. The strong localization of the continum, and its nonthermal character, must be contrasted to the thermal continuum of galaxies due to the stars forming them.

3. The fact, recently discovered by repeated measurements, that both radio and optical luminosities of at least some of the quasars are variable in time.

From these two contrasting aspects, two lines of thought have been derived, one stressing the similarities, the other the discrepancies between quasars and galaxies. We shall now discuss their arguments and consequences.

A first important result obtained along the first line has been the discovery of the so-called QSO or QSG (quasistellar objects or quasistellar galaxies), whose existence has been argued using the following point of view: Strong radio galaxies are a kind of exception among normal galaxies, due to a momentary state of explosion. If quasars are to be compared to radio galaxies, should it not be necessary to suppose that they also are representing an exceptional exploding state of an up to now unknown population of quiescent quasistellar objects, that is, concentrated structures on a galactic scale, which should normally be nonradio emitting and could be called normal QSO's, in contrast to the radio emitting ones, which are the QSS up to now detected?

Two years ago, Sandage[2,3] was thought to have identified such QSO's. He sought to determine the number of pointlike objects (mostly stars) of blue color as a function of their apparent magnitude in the direction perpendicular to the plane of our galaxy. At magnitude 14.5 he found a break both in the slope of the curve and in the color, the fainter stars having an ultraviolet excess ane the higher slope corresponding to a uniform distribution in space. Sandage thought that the stars of the galaxy were ending at magnitude 14.5 and that the fainter objects seen beyond this limit were just the extragalactic QSO's looked for. In some cases, the measurement of a strong red shift substantiated this interpretation. Assuming all the faint objects to be QSO's, Sandage gave for the ratio $N(QSS)/N(QSO)$ the figure 1/500.

However, the quantitative, not the qualitative, conclusions of Sandage turned out to be wrong, as some other authors were able to show that a great part of the supposed QSO's of Sandage were just those peculiar stars with ultraviolet excess (novae, white dwarfs) which we have mentioned earlier and which Sandage neglected in his considerations. However, although the frequency of QSO's is certainly lower than the value assigned by Sandage, the qualitative aspects of his discovery remain true, as the objects with large red shift cannot be interpreted differently. One may therefore conclude that quiescent nonradioemitting quasistellar objects exist as a normal background of the QSS themselves, although their true frequency is not yet well established.

Meanwhile, on the opposite side, the line stressing the contrasts between quasars and galaxies was apparently based upon an argument which, as already mentioned, appeared to grow stronger and stronger with the gradual recognition that several quasars were varying in time in their energy output, both on radio and optical frequencies, among them some of the best known objects such as 3C 273 and 3C 48. Three different types of variations were observed: periodic ones amounting to few tenths of magnitude with a period of about 10 years; rapid flares up to 1 magnitude amplitude, lasting some days or weeks; irregular fadings of the light intensity down to a factor two, occurring in a few months time. Whereas flares could be thought as due to local outbursts in the quasar, it was difficult to explain long rate decreases of light in a relatively short time for objects on a galactic scale. In fact, on general grounds, it is expected that an object whose light output varies in a time $\tau$ cannot have dimensions greater than $c\tau$, and the $c\tau$ values obtained from the quasars were much smaller than their dimensions deduced from the cosmological assumption concerning their distances.

Therefore, to explain variability, it seemed necessary to reduce all the dimensions of the quasars; and to that aim, near them; and therefore interpret the red shifts as being not cosmological.

However, if one still relied on the exclusion of a gravitational effect, the only possibility was to think of the QSS as being thrown at high velocities by some powerful explosion. And an available source for such explosions were just the radio galaxies. Now in order that all shifts should be red, one had to assume that they were thrown either by our own galaxy, or by a nearby one, such as Cen A, giving signs of having exploded not too long ago. Both assumptions were made and discussed by Terrell[4] and by Hoyle and Burbidge.[5] Assuming that our Galaxy exploded $10^7$ years ago, the maximum distance reached now by the QSS was found to be of the order of 1 Mpc, and in the case of Cen A, a bit greater, namely, 10 Mpc. In any case, these were quite "local" distances with respect to the cosmological ones; hence, the name of local theory given to this approach. On such a new assumption, it was then shown that on the whole all quantities related to quasars as mass and diameter were smaller by a factor between 100 or 1000 with respect to those for the cosmological model. In this way the objection concerning variability was settled and no contradiction was any more at hand.

Moreover, it was argued that if Cen A should have exploded,

then some particular quasar shot just toward us could have shown a blue shift. And it was therefore pointed out by Hoyle and Burbidge that should we observe even a single blue shift, the local theory would be confirmed, while the lack of such an observation would not invalidate it, as the solid angle probability of a quasar being shot just in our direction was exceedingly small.

The strongest objection to this "restricted" local theory was, of course, the assumption that a single galaxy was exploding and responsible for shooting all the quasars. It should have appeared much more likely to think of a uniform distribution of galaxies exploding through space, but then, why should all shifts have been red?

The test for this second kind of assumption, or "extended" local theory, had to be provided by the calculation of the ratio of blue to red shifts to be expected. However, it turned out that this ratio was very sensitive to the specific assumptions made, and lead to quite different predictions according to their choice. In fact, if one calculated, as done by Zapolsky,[6] the ratio according to purely geometrical grounds based on solid angle considerations, the ratio number of blue/number of red shifts was smaller than one and approaching zero for high values of the velocity of the quasars. If, however, the same probability was derived with the further assumption of considering only objects with the same apparent luminosity, as done by Faulkner and co-workers,[7] the result was just the opposite—blue shifts exceding by far the red ones with increasing velocity. The comparison of these calculations with data then required an accurate discussion on the selection criteria of the material of observation, which does not seem to have been done, although to my feeling the available selected objects fulfill Faulkner's rather than Zapolski's assumption, and if this were true, local theory with widespread exploding sources should be directly disproved.

Another point which has been looked for, in deciding between the local and the cosmological approaches, is related to possible absorption lines to be found in the spectra of quasars, which, if due to intergalactic material, should, by their absence of red shift and by their strength, give us indications on the real distances of the QSS. Here also, up to now, no definite conclusions have been reached, the most significant limitation thus far obtained being due to an absorption line observed in the spectrum of 3C 273 which is attributed to an hydrogen cloud of the Virgo Cluster, nearby which 3C 273 happens to be situated. The distance of the Virgo Cluster being 11 Mpc, we thus get this figure as the minimum distance for 3C 273, which is about just the upper limit postulated by the local theory.

However, although no decisive proof thus far has been reached to allow one to decide between the two possible scales for quasar phenomena, the tide is actually turning against the local approach. This is because, now as gradually shown in several papers, it gives rise to several contradictions, which on the whole make the local model even less adequate to explain facts and more inconsistent than the cosmological one, owing to a lack of a really significant picture explaining the physical mechanism and the energy balance of the explosions responsible for throwing such enormous masses such as quasars ($10^6 \sim 10^7$ M. even in the local theory) at such incredibly high speeds as is required by their red shifts. Therefore, the most recent approaches to the problem are mostly intended to remove the objections to the cosmological description by the collection of more refined data and more elaborate discussions of them. In what follows, we shall briefly try to summarize the points thus reached and the situation following from them.

First of all, more detailed radio spectra have been obtained by studying both the correlations of the shape differences which have been mentioned at the beginning with several other characteristics, and the changes in time related to the luminosity variations. It has thus turned out, according to the results of Dent and Haddock,[8] that there is a definite relation between the shape of the spectrum and the distance of the radio sources from the optical image. If sources are generally expanding, as is believed from all the evidence, then flat spectra, or those with an upward convex curvature, belong to young sources at the beginning of the expansion. As the source grows in extension, the maximum of the spectrum is shifted toward lower frequencies, according to the laws of synchrotron self-absorption, and the slope of the spectrum becomes more steep and more straight. For what concerns the upward concave spectra, an analysis of 3C 273 through lunar occultation has revealed that the curve may be divided into two components, one of them, emitted by the jet, being a pure $\nu^{-\alpha}$ power curve, while the other one, emitted by the point source itself, is practically flat at high radio frequencies and typical of a source at its earlier stages of expansion. Therefore complex cases may be generally thought as composite situations, representing the outcome of successive explosions, the older outer shells emitting straight power spectra whereas the younger ones, much more concentrated, radiate a self absorbed spectrum which is more important on the high-frequency side.

These conspicuous results have been completed by the detailed time variation analysis of the spectra of some well-known sources (3C 273, 3C 279, and 3C 345) by Moffet.[9] It has thus turned out that the time variation is confined only to the higher emission frequencies of the spectrum, while the lower ones are not changed. According to the previous analysis, this fact is a clear indication that the variable part is confined to the inner regions of the source and affects only the newly young expanding shells. This, in effect, as may be immediately understood, just meets the contradiction between rapid light variations and cosmological dimensions if, as it turns out, light variations are just confined to a small part, the innermost one, of the source. Further studies now seem to point out that, on this main line, light variations may be understood and explained, so that, although it is certainly too early to say that the whole aspect of the situation has been solved, one may at least admit that there are no more strong objections in principle to the cosmological approach based on this evidence. Therefore, an understanding of the phenomena observed may be hoped to be reached when further more extended and systematic measurements will be available.

If, according to the preceding results, the cosmological interpretation of quasars appears to be no more incompatible with the experimental data at hand and, further, the less troublesome explanation in the present situation of the whole phenomenology collected, the next step that may be immediately tried, owing to its fundamental importance, is to use quasars as cosmological objects in order to show the depths of our universe. If, as was already stated, quasars are on the average 100 time more luminous that the brightest radio galaxies, they will allow us to observe space at distances about ten times greater than before. And, therefore, they will enable us to look far back into the past, since information coming to us from $10^{10}$ light years distances reveals us things as they looked $10^{10}$ years ago.

As is well known, the two main modern lines of interpretation of the structure of our universe which are adequate to explain all important data we now have, are given by the following: the evolutionary cosmologies, which interpret the red shift of the galaxies as an expansion of the universe from an initial concentrated state and which by integrating the equation of general relativity are able to produce several types of models, all of which have the same common feature but with several departures from Euclidian geometry; and the steady-state theory, rooted on the assumption of the perfect cosmological principle, which asserts that the universe must look exactly similar to any observer wherever located either in space or in time. This approach arrives at the conclusion that the universe must be Euclidian, but since it is expanding and since its mean density should remain the same, in order to satisfy the cosmological principle, matter must be continuously created, the rate of creation required being, however, so small as to escape the actual experimental possibilities of detection. The main differences descending from these different points of view are the laws giving the number $N$ of galaxies as a function of their apparent luminosity $S$, and the connection between apparent luminosity and red shift, which both depend upon the structure of space. These laws, however, have the same form for all cosmologies when distances are small, and differences become significant only from 1000 Mpc upward. As we have seen, most observed galaxies are not located so far, and for them in fact the $N/S$ counts were just indicating Euclidian structure of space, but quasars are, on the main, much more distant and, therefore, may be expected to give some answer on this most fundamental problem of the structure of the universe.

The $N/S$ law has been tested by different authors, mainly Veron[10] and Bolton,[11] and the results are that the slope of the curve log $N/$log $S$, while being 1.5 for galaxies as expected for an Euclidian space, becomes for quasars and unidentified sources 1.85, thus revealing a definite discrepancy with the Euclidian model and disproving the steady-state cosmology. Moreover, the relation between luminosity and red shift enables us to go a step further, and indicate whether this excess of sources at great distances is due to purely increased density, or also increased luminosity. It appears thus far that experimental data are best interpreted by this second assumption and that in the remote past one should think quasars and galaxies as having been both more luminous and concentrated in space than they are now. Of course, these deductions are to be taken actually as pure indications and not yet as definite results.

After having thus reviewed both the data leading to a consistent understanding of the model for quasars and the consequences that may be drawn from the assumption of those objects being cosmological, let me as a last point briefly summarize the physical views concerning the cause of the explosions giving rise to the quasars and to the tremendous energy output they require, which, perhaps of all the unknowns of the problem, have been up to now the most mysterious and most hypothetized.

The first ideas expressed on this subject are common for quasars and radio galaxies, and have been tried, as is usual in physics, to explain new observed events on the basis of facts already familiar or easily guessed. As such attempts have to be viewed both the supernova and the star collision theories. The first assumes that the energy output of nuclei of galaxies or quasars is due to the simultaneous outburst of a large amount of supernovae or stars exploding at the end of their evolution. It is however not clear why all these explosions should occur all together, as they are generally supposed to be intrinsic phenomena for any single star triggered by its internal disequilibrium conditions. From this point of view the multiple star collision assumption due to a very-high density in the galactic nuclei is a somewhat better model, as it can in fact be shown on dynamical grounds that star density may be expected to increase during the life of a galaxy, and should it succeed to reach a given critical value, collisions could become so frequent as to give rise to the observed explosive phenomena. The last step on this line of thought may be considered the superstar theory, namely, the formation of a single enormous superstar appearing to us as a quasar with mass of the order of $10^8$ solar masses in the center of a galaxy as a consequence of its extreme condensation. Fowler has developed in detail such a model and has been able to show that it should be compatible with part of the facts observed.

However, there are also some other ideas, far less conventional, which, although they have to be taken with care and only as possible working hypothesis, may at least suggest that the quasar phenomenon could provide a test for some of the most unexpected consequences and the less intuitive fields of theoretical physics, such as general relativity and elementary particles phenomenology. And I would like to paraphrase, in conclusion, an idea of Novikov and Ne'eman,[12] which may be considered as typical of this kind of approach. If we assume according to evolutionary models that in its early stages the universe was concentrated in an extremely small volume, then the density of matter should have been so high that masses could have been compressed into their Schwarzschild radius, and therefore escape observability from the outside. When expansion set in, the density gradually decreased, masses expanded outside the Schwarzschild radius, and became gradually visible. Now, if density fluctuations were at hand in the initial stage, the most dense parts of the universe were kept longer inside their Schwarzschild radius, and may be that still some of them only now are coming out from its limits. It is these delayed portions of superdense matter just now becoming detectable that are supposed to be observed as quasars. And if one wonders why this apparition from unobservability occurs with such a tremendous output of energy, well, the answer is at hand and provided by elementary particle data. In the highly compressed state of the beginning, density is so highly supernuclear that, owing to the Pauli principle acting on the fermionic decay products, the stable states of baryons are neither the proton nor even the neutron but rather the higher-mass hyperons of which we know today a whole range. As soon as density decreases, hyperons begin gradually to decay into nucleons, and it is therefore their mass excess with respect to nucleons that is irradiated and provides the tremendous output of light and radio waves we observe.

This last suggestion is aimed at showing that quasars will perhaps prove to be a good test for the most advanced fields of theoretical physics. However, data are being gathered so quickly and the situation is so rapidly changing that the present summary may be completely out of date in the next few months. In conclusion, it is therefore wise to stress that most of what has been reported is still to be considered as highly speculative and liable to be radically changed.

## REFERENCES

1. J. L. Greenstein and M. Schmidt, *Astrophys. J.* **140:** 1 (1964).
2. A. Sandage, *Astrophys. J.* **141:** 1560 (1965).
3. A. Sandage and P. Veron, *Astrophys. J.* **142:** 412 (1965).
4. J. Terrell, *Science* **145:** 918 (1964).
5. F. Hoyle and G. Burbidge, *Astrophys. J.* **144:** 534 (1966).
6. H. Zapolsky, *Science* **153:** 635 (1966).
7. J. Faulkner, J. E. Gunn, and B. A. Peterson, *Nature* **211:** 502 (1966).
8. W. A. Dent and F. T. Haddock, *Astrophys. J.* **144:** 568 (1966).
9. A. Moffet, preprint n° 6, Owens Valley Radio Observatory, Caltech, Pasadena, 1966.
10. P. Veron, *Nature* **211:** 724 (1966).
11. J. G. Bolton, *Nature* **211:** 917 (1966).
12. Y. Ne'eman, *Astrophys. J.* **141:** 1303 (1965); I. D. Novikov, *Astr. Zh.* **41:** 1075(1965) [Soviet Astr. AJ 8, 857 (1965)].

# On the Unification of Physical Theories

ANDRE MERCIER

*BERN UNIVERSITY‡*
*Bern, Switzerland*

From antiquity down to the Renaissance, "theories" advanced for the explanation of the behavior of nature were based on the assumption of existing harmonies—harmonies of numbers, spheres, and other geometrical figures. Kepler was still of that opinion when he wrote his famous *Harmonices mundi*. All such attempts can be considered as aiming at a unifying view of the world of physical (especially celestial) bodies and systems.

When Newton expounded his mechanics, a completely different view was adopted. Of course, his prime interest was directed toward the motion of celestial bodies, and he succeeded in explaining it, not on the basis of pre-established harmonies but on the assumption of a system of dynamical causes. He also succeeded in explaining both the motion of celestial bodies and the motion of bodies falling onto the surface of the earth as being due to one and the same force. This unification of the two forces, which till then must have appeared totally disparate, was a sensational achievement. The question immediately arose: Could other forces of nature also be reduced, like the above, to a unique force? The answer was no. Thus, there was a setback at least for the time being in the program of unifying forces of nature. Yet, Newtonian mechanics was found to be useful for studying theoretically any question about the physical world. Not only general theorems (concerning mostly conservation laws),

‡Department of Theoretical Physics.

established independently of the specific force at hand, but also any law of force could also be introduced into the fundamental equations; thus mechanics came to be the universal theory according to which all (lifeless) nature seemed to be working. This is what may be called the unification of the theoretical model of the physical world. This does not, however, imply the unification of the possible laws of nature—like gravitation, friction and, later, electromagnetism, etc.—in short, what are today called the various interactions.

In spite of a formal similarity, Coulomb's law is not reducible to gravitation. In the course of the 19th century, it gradually appeared that there were (at least) two irreducible interactions: gravitation and electromagnetism. At the beginning, electricity and magnetism were separate phenomena, and separate laws were looked for. We know today that since electricity is related to the presence of charge in space and magnetism to the motion of charge, these two phenomena are only two aspects of one and the same interaction. Their unification is to be considered a sensation as great as when gravity was found to be the same as gravitation.

The phenomenon of light had long resisted attempts to explain it in either mechanical terms (Newton's light corpuscles) or according to the wave theory (Huygens and Fresnel) with an assumption as to the existence of some quasielastic ether, until Maxwell succeeded in reducing it to electromagnetism—again, a new sensation in the sense of a unification. However, this belief of Maxwell in the existence of an ether capable of stresses, was later disproved.

Thus at the end of the 19th century, only electromagnetism and gravitation seemed to be the forces to be reckoned with as independent interactions. The only unifying feature is the Lagrangian or Hamiltonian formulation employed in these theories. Even here it is to be noted that electromagnetism needed the field concept while the Newtonian theory of gravitation involved action at a distance.

Even earlier, no attempt was made to understand the elastic forces and their relation to the forces of gravitation. It was believed that the former could be disposed of by means of certain constants depending on the material.

In the course of the development of physics, scientists did not have the same understanding of these matters as we do now. It was Albert Einstein who clarified the situation. Realizing that special and (later on) general relativity theory are unable to deal satisfactorily with both interactions at once, he set out to construct what has since been known as a unified (field) theory. Neither Einstein, nor any one else actually succeeded in elaborating a theory that would satisfy all possible criteria. Therefore, the Einsteinian dream may be considered the tragic side of his scientific career. Nevertheless, there is, beside the grandeur of his attempts, the development of a new trend in physics to unify things that appear to be separate.

When the special theory of relativity was put forward, it was a great blow to Newtonian mechanics. Until then, it had held the premier position of a theory, apparently suited for any law of force (any "reasonable" Lagrangian, Hamiltonian, etc.). But Newton's law of gravitation, not being Lorentz-invariant, had to be dropped from special relativity theory; and Newtonian mechanics, working within the Galilean conception of relativity, was found to be in fact suited only for dealing with gravitation. So it was reduced to a mere gravitodynamics.

Einstein attempted a double generalization in getting rid of the special frames of reference and in reintroducting gravitation by means of the geometry with a Riemannian metric; a most remarkable step indeed. However, something very disappointing happened. If gravitation was beautifully explained, including even new effects, electromagnetism was not because Maxwell's equations do not follow from Einstein's field equations and because the procedure consisting of artificially making them generally covariant in order to glue them onto general relativity is very unsatisfactory. "General" relativity has to be generalized, if the two interactions are to be unified. Attempts have been made either by enlarging the number of dimensions of the manifold (adding some rather arbitrary conditions of cylindricity), or by replacing the Riemannian picture by some other picture (Finslerian picture, or a variety with nonsymmetric affine connection). But none of the attempts have led to satisfactory results.

One gets the impression that gravitation and electromagnetism resist a unification of interactions; possibly they need separate theories for their explanation! This would be a very serious matter for epistemology.

From the point of view of the various interactions, we notice that if an interaction constant is used as a measure of the intrinsic intensity of the interaction, then such constants must be used in all laws of interaction. It would be physically meaningless not to do so. Therefore, if gravitation is written as a force $-G(m_1 m_2/r^2)$, then Maxwell's equation must contain a corresponding constant, $\epsilon_0$ say, according to international recommendation of the SUN-Commission of IUPAP; however, if the magnetic induction field is gauged by a factor $c$ (and $\epsilon_0 \mu_0 = c^{-2}$), the most elegant and, to our mind, most meaningful form of these equations will be (for the so-called vacuum)‡

$$\epsilon_0 \nabla \cdot \mathbf{E} = \rho \qquad \epsilon_0 \left( \nabla \times \mathbf{B} - \frac{\partial \mathbf{E}}{c \partial t} \right) = \rho \frac{\mathbf{v}}{c}$$

$$\nabla \cdot \mathbf{B} = 0 \qquad \nabla \times \mathbf{E} + \frac{\partial \mathbf{B}}{c \partial t} = 0$$

$$\mathbf{F} = e \left( \mathbf{E} + \frac{\mathbf{v}}{c} \times \mathbf{B} \right)$$

from which follows, for the energy-density, Poynting's vector and Maxwell's stress-tensor expressions containing $\epsilon_0$ as a factor:

$$\mathcal{E} = \frac{\epsilon_0}{2} (\mathbf{E}^2 + \mathbf{B}^2) \qquad \mathbf{S} = \epsilon_0 c \, \mathbf{E} \times \mathbf{B}$$

This is most satisfactory, for energy-momentum magnitudes should indeed be determined by the strength of the interaction. As a matter of fact, interaction constants are the most important quantities to be experimentally determined as accurately as possible, for they are the indications of how strong the physical systems are held together in nature. It would, therefore, appear unreasonable to set them equal to 1, as is often done for so-called practical reasons, for they do not possess the nature of standard units since no standard of an interaction constant can be artificially constructed and deposited in a bureau of standards at Sèvres or elsewhere. Certain other universal constants like $c, \hbar, \ldots$ can reasonably and should indeed be taken as theoretical units, for it is in *principle* possible to construct standards of velocity, action, etc.

Now, in consonance with this remark, the Giorgi system of physical dimensions should be used for the combination of gravitodynamics with electrodynamics. If further independent interactions are discovered, each one will require a new interaction constant, and correspondingly we may expect the Giorgi procedure of enlarg-

---

‡See A. Mercier, "Leçons sur les principes de l'electrodynamique classique," (Neuchâtel, Ed. du Griffon, 1952).

ing the independent physical dimensions to apply. Many a theoretical physicist will shake his head at this, but we ask him to consider the matter with care, for it is connected with the possibility or impossibility of unification of interactions. Sommerfeld, after defining the fine structure constant as equal to $e^2/\hbar c$, later changed it to $\alpha = e^2/4\pi\epsilon_0\hbar c$. Pure numbers of this kind, sometimes called Eddington numbers, will all have to contain one specific interaction constant or no pure number can be obtained. Only then will such numbers have a meaning as natural intensities of interactions.

In connection with these remarks, we may raise the question of the origin of mass put. According to Poincaré, the mass of an electron can be explained in terms of its electric charge. This was generalized in quantum theory to the notion of self-energy. But such self-energies arise from every coupling allowed in the quantum theory of fields, so mass might be a complex function of the quanta of various fields; if it diverges, renormalization is at hand, though this is rather artificial.

The question also arises whether it is conceivable that one and the same law of interaction contains more than one interaction constant. The answer can be given by considering the Einstein field equations. As is well known, they amount to a proportionality between a tensor $C$ and the tensor $M$ of energy-momentum:

$$C = \chi M$$

where $C$ is constructed from the tensor of Riemann–Christoffel and $\chi$ is called Einstein's constant. Whereas Einstein chose for $C$ a simplified combination of derivatives of the $g_{ik}$'s, Cartan showed that it was a special choice and gave the most general form which $C$ may take (Cartan's tensor) such that, apart from singularities, the vanishing of $C$ would imply a flat space–time. $C$ includes one more constant $\lambda$, called the cosmological constant. Therefore, there are two possible constants in Einstein's theory of general relativity. The presence of both $\chi$ (which is found simply related to $G$ and $c$) and $\lambda$, is still rather mysterious. But one gets the impression, that "two" interactions are depicted in "one" law of interaction. This will have some bearing on our concluding remarks.

Now suppose one could succeed in unifying gravitation with electromagnetism. As is done in the quantum theory of fields, one might try to quantize the unified theory. Of course, a quantization of that sort will prove awkward, if at all possible, because Einstein's field is already difficult to quantize, since it is the solution of non-linear equations. But in the linear approximation, one gets equations typical of a spin-2 particle; therefore it is said that the quanta of the gravitation field, or gravitons, are such particles. On the other hand, the quanta of the electromagnetic field are photons. How is it then, if the unification succeeds, that one does not know not only of one unique "ton," but also of two widely different things with different properties?

We may mention in passing that if asked whether two gravitons interact by gravitation, a specialist will most probably answer yes—a most curious phenomenon, to be sure.

Since we are now using quantum-theoretical considerations, let us ask: What kind of theory is quantum theory? According to the point of view adopted this question admits of two answers.

From the point of view of interaction, quantum theory is a theory of electromagnetism. One sees it already in classical quantum theory even in the case of a free particle for which the charge need not be explicitly taken into account. For, given the dynamical four-vector $(E, c\mathbf{p})$ of energy-momentum of a particle and the wavelike four-vector $(v, c\mathbf{x})$ of frequency-wave number of a corresponding wave, the only reasonable connection between both is that of de Broglie,

$$(E, c\mathbf{p}) = h (v, c\mathbf{x})$$

since space–time as such cannot be but isotropic and homogeneous. Therefore, de Broglie's material waves rest on special relativity, which is applicable only to electrodynamics. The fact that Schrö-dinger's wave mechanics was then applied to nonrelativistic problems does not change this conclusion; it was done for practical reasons. Atoms, molecules, and crystals are all what they are because of the electromagnetic interaction, and even exchange forces and Dirac's original spinning particles are of purely electro-magnetic nature.

It may be asked whether (ordinary, not isotopic) spin is or is not a purely electromagnetic manifestation. In any case, we know that spinors do form a representation of the Lorentz group, and what is the Lorentz group physically connected with, if not with electromagnetism?

Did not Dirac conceive of quantum theory as of a theory of (electromagnetic) radiation? Of course, quantum theory in its more recent form of quantum electrodynamics is obviously a theory of electromagnetism.

From the point of view of its machinery, quantum theory was found to work as a universal theory as did classical mechanics before it. This is not surprising considering its character as a canonical formalism. Any Hamiltonian operator (satisfying reasonable conditions like Hermiticity) can be used in Schrödinger's equation. In the quantum theory of fields one can introduce as many Lagrangian densities as one wishes, with or without charge and with various tensorial properties; it seems as if nature had taken advantage of this, since the various fields seem to be actually realized in nature.

However, from this last point of view referring to quantum fields, the theory is meant essentially for free fields. Interactions are not directly considered as such; rather, coupling between different kinds of fields are used. These couplings involve coupling constants.

Thus, quantum theory also works for nonelectromagnetic phenomena. Or rather, it seems to do so. But in view of the reduction of classical mechanics from a universal theory to a mere gravitodynamics, we should be very careful in our statement and be ready at some future date to find quantum theory suited only for quantum electrodynamics, especially since almost every one expects that some super-theory will be found, of which quantum theory will only be some sort of approximation.

If it is not quite absurd to ask if gravitation can be quantized, neither is it to ask if quantum (field) theory can be made generally covariant. However, we may doubt whether these questions make much sense. Firstly, if a unification of gravitation with either electromagnetism or with any other interaction cannot take place on the mere basis of a Riemannian manifold, what would the "general" covariance of these other fields consist of? This question is connected with the statements repeatedly made by V. Fock, according to whom one should be cautious about requirements of general covariance. Though the latter is mathematically well-defined, what is physically interesting is to know how and what is measured. Secondly, since quantization, at least in the usual sense, is an operation to be done on superposable fields, and since there is no reason to confine ourselves to the linear approximation in gravitation since there may be strong deviations in the $g_{ik}$'s from those of a flat space, what is the correct way to quantize gravitation, if at all?

This last contention does not by any means imply that gravitational measurements escape complementarity. It is not the purpose of this paper to explain the reason. L. Rosenfeld has done so at the symposium of the Berlin Academy, November 1965.

But complementarity belongs to the class of connections between theories which makes such theories, constructed for seemingly different purposes and interactions, not completely separable from one another. General theorems (conservation laws) belong to the same class. Even if Einstein's gravitodynamics (ordinary general relativity theory) should discard electromagnetism as an interaction (and must do so on account of its skew symmetry which can never be expressed either in terms of the $g_{ik}$'s or the Riemannian affine connection), it cannot totally ignore it for electromagnetic energy-momentum tensor contributes to $M$. This tensor is, by the way, the only quantity which is totally symmetric in electromagnetism.

This necessity is similar to, though not equivalent with, the difficulty of getting rid of the ratio of charge to mass. Whether charges (and quanta of other fields) exist independently of mass is a question that may have relevance to unification. No theory has yet been made with nonmassive particles (zero rest mass but particle-velocity $c$ is not a nonmassive case), but one might conceive of a theory that would differ from the usual ones.

It seems as if particles could be classified in a hierarchy of groups according to main mass of the group representative. (Group here does not mean a mathematical one). There would be a series of such groups. (Mass differences within a group may be explained along the line of a symmetry violation.) If this series would close itself in a circle, this would mean that each coupling by pair of fields is but one limb of a chain, and this would realize the unification from the point of view of interaction. If it does not, one does not see clearly why there should be only certain couplings. These matters are not easy to discuss, since one has no precise data and since our theories are not yet sufficiently advanced.

To summarize what has been considered thus far, we might say that there are three ways of achieving unification of physical theories in the future, each with defects which will make its case a weak one.

1. One can conceive of a key-dynamics, as classical mechanics

or quantum theory has been, allowing for any reasonable law of interaction, to be used in some postulated fundamental equation.

However, this is not the unification of interaction as dreamed of by Einstein. Moreover, one sees no physical reason why certain laws (gravitation, electromagnetism, weak and strong interactions, etc.) should be realized in this world and not others.

2. One can imagine a closed dynamics, in which a unique formula, e.g., of the form

$$\delta \int_\omega L \, d\omega = 0$$

would contain a unique, universal form for $L$, namely, $L_\omega$, explaining at the same time all specific interactions as they are seen separately. This $L_\omega$ would have to include corresponding interaction constants, say, $n$ constants.

This is the kind of theory Einstein was aiming at. With Cartan's tensor, already two constants are at hand. However, one does not see why there should be just $n$, not $(n + p)$, such constants. This approach is subject to the same criticism as under (1).

Moreover, and this is more serious perhaps, case (2) would have a meaning only if $L_\omega$ is unique. And then the question arises: How is it that the Creator made the world like that? The only answer to this question is the Leibnitzian answer: The world is the best of all possible worlds. This leads to theology.

3. It might be that the unification would avoid a dynamical foundation altogether: What do we have then? Perhaps just a set of numbers like the fine structure constant. This would be a Platonism like that dreamed of by Eddington. Also, why should just this set of numbers be important and not the totality of mathematics?

We have no conclusion to offer. Each time a scientist believes he has made a unification of theories, he has not really done so. For nature will reveal to him new features which do not enter nicely into his system. This does not deter work on further attempt toward unification. But it shows that a final unification is utopian (at least we think so) and this is good, for otherwise there would be nothing left to do.

# A Note on the Representations of Dirac Groups

ALLADI RAMAKRISHNAN AND
I. V. V. RAGHAVACHARAYULU

*MATSCIENCE*
*Madras, India*

## 1. INTRODUCTION

In a recent paper,[1] a hierarchy of matrices $L_m$, which contains the Dirac Hamiltonian as a particular case, ($m$ represents the number of parameters occuring in $L_m$) was introduced. These matrices can be expressed as linear combinations of matrix representations of Clifford elements[2] satisfying anticommutation relations, the parameters being the coefficients. In obtaining the hierarchy of matrices $L_m$ in a systematic way, a $\sigma$-operation is defined which corresponds to the introduction of two additional parameters.

In this paper, we shall study the group theoretical significance of the hierarchy of matrices and the $\sigma$-operation by introducing a group called Dirac group $G(m)$. For that, given $m$ Clifford elements $\{\gamma_m\}$, hereafter called Dirac operators, augment $\{\gamma_m\}$ by two more elements $E$ and $\bar{E}$. Now making use of $\{E, \bar{E}, \{\gamma_m\}\}$, as generating elements, we set up the group [3,5-7] $G(m)$ whose generating relations are $\gamma_i \gamma_j = -I\gamma_j\gamma_i$ and $-1 = \bar{E}$. Now obviously the well-known theory of group representations can be used in setting up the matrix representations of Clifford elements satisfying the usual anticommutation relations through $G(m)$. In particular, we make use of the Mackey theory[4] of induced representations in setting up the representations of the Dirac group $G(m)$. Since the Dirac group is solvable,[3] the Mackay theory reduces to the little group technique and enables us to obtain directly all the irreducible representations of the group in steps. The $\sigma$-operation is found to be identical with this process.

## 2. THE DIRAC GROUP

As in Ref. 1, let us introduce a hierarchy of square matrices $L_m$ involving $m$ independent continuous parameters $\lambda_1, \lambda_2, \ldots, \lambda_m$ such that

$$L_m^2 = (\lambda_1^2 + \lambda_2^2 + \ldots + \lambda_m^2)I \qquad (1)$$

Obviously, $L_m$ should be linear in each one of the parameters $\lambda_i$, and can be taken as

$$L_m = \mathscr{L}_1\lambda_1 + \mathscr{L}_2\lambda_2 + \ldots + \mathscr{L}_m\lambda_m \qquad (2)$$

where $\mathscr{L}_i$ are the "generator matrices" independent of the parameters. Note that the dimension of the matrices $\mathscr{L}_i$ is not yet specified. Now imposing condition (1), we note that the $\mathscr{L}_i$ satisfy the anticommutation relations

$$(\mathscr{L}_i\mathscr{L}_j + \mathscr{L}_j\mathscr{L}_i) = 2\delta_{ij} \qquad (3)$$

From the nature of relation (3) it is obvious that $\mathscr{L}_i$ can be looked upon as $m$ Clifford elements $\gamma_i$, and from (1) and (2) it is obvious that we are considering their matrix representations. If $m = 4$, these $\mathscr{L}_i$ are known as Dirac matrices. Hence, hereafter, we shall call $\mathscr{L}_i$, in the general case also, Dirac matrices, which when irreducible, are matrix representations of the Clifford elements. (*Hence, Dirac matrices are matrix representation of the Clifford elements for the case $m = 4$.*) To motivate the study of the Clifford elements in the general case when $m \neq 4$ we give another instance[7] from physics where they are of importance.

In the quantum field theory, one frequently encounters a set of operators $a_1, a_2, \ldots, a_n; b_1, b_2, \ldots, b_n$ such that

$$\begin{aligned} [a_i, a_j]_+ = [b_i, b_j]_+ &= 0 \\ [a_i, b_j]_+ &= \delta_{ij} \end{aligned} \qquad (4)$$

where $a$'s and $b$'s are called annihilation and creation operators. If we now form the operators $q_i = a_i + b_i$ and $p_i = -i(a_i - b_i)$ then we have

$$[q_i, q_j]_+ = [p_i, p_j]_+ = 2\delta_{ij}$$
$$[q_i, p_j]_+ = 0 \tag{5}$$

Now, let $q_i = \gamma_{2i-1}$, $p_i = \gamma_{2i}$. Then collectively $\{\gamma_i\}$, where $i = 1, 2,$ $\ldots, 2n$ satisfy the anticommutation relations (3). Hence, the study of the representation theory of Clifford elements satisfying anti-commutation relations is of basic importance in physics.

In Ref. 1, the representation matrices of the Clifford elements are obtained by matrix methods in a systematic way by introducing a $\sigma$-operation. This note has motivated us to study in detail the mathematical significance of the $\sigma$-operation.

Consider the set $\{\gamma_i\}$ of $m$ Clifford elements and augment them by $E$ and $\bar{E}$. Now define a group with the generating elements $\{E, \bar{E}, \{\gamma_i\}\}$ satisfying the following generating relations:

$$\gamma_i \gamma_i = \bar{E}^2 = E$$
$$\gamma_i \gamma_j = \bar{E}\gamma_j \gamma_i$$

and

$$\bar{E}\gamma_i = \gamma_i \bar{E}$$

Obviously, $E$ is the identity element of $G(m)$ and as the order of each element is 2, all the elements of the group are given by $\bar{E}^{j_0}\gamma_1^{j_1}\gamma_2^{j_2}\ldots\gamma_m^{j_m}$, where each $j$ is either zero or unity. Its order is $2^{m+1}$ When $m = 4$, the $\gamma_i(= \mathscr{L})$ correspond to the Dirac matrices. When $\bar{E} = -1$, we call $G(m)$ the Dirac group.

Now, $G(m)$ has $1 + 2^m$ when $m$ is even, or $2 + 2^m$ when $m$ is odd, conjugate classes given by $E, \bar{E}, (A, \bar{E}A)$ for all $A \in G(m)$ and $A \neq E, \bar{E}$ when $m$ is even $E, \bar{E}, \gamma_1 \gamma_2 \ldots \gamma_m(=\delta_m), \bar{E}\delta_m, (A, \bar{E}A)$ for all $A \in G(m)$ and $A \neq E, \bar{E}, \delta_m, \bar{E}\delta_m$ when $m$ is odd, respectively.

Further, we can very easily establish the following properties regarding $G(m)$:

1. Every subgroup of $G(m)$ different from $E$ contains the normal subgroup $G_0 = \{E, \bar{E}\}$ and, hence, $G_0$ is the minimal normal subgroup.

2. $G(p)$ is the proper maximal normal subgroup of $G(p + 1)$ and $G(p + 1)/G(p)$ is a factor group of order 2. Hence, considering the composition series

$$G(m) \supset G(m - 1)\ldots \supset G_0 \supset E$$

where each factor group is of order 2, it follows that $G(m)$ is a solvable group.


## 4. REPRESENTATIONS OF THE DIRAC GROUP

To set up the matrix representations of $G(m)$ let us apply the Mackey technique of induced representations to reduce $G(m)$ with respect to the normal subgroup $G_0$. The one-dimensional irreducible representations of $G_0$ are given by $\Gamma_\pm$: $\bar{E} \to \pm 1$ when specified through the generating element $\bar{E}$ of $G_0$. The orbits of the representations of $G_0$ relative to $G(m)$ are $\Gamma_+$ and $\Gamma_-$, and their stability groups are the same and are given by $G(m)$. So the representations of $G(m)$ fall into two classes: those in which $\bar{E} \to I$ and $\bar{E} \to -I$ as the element $\bar{E}$ commutes with all the elements of $G(m)$ and $\bar{E}^2 = E$.

*Class 1*: Consider the induced representation $\Gamma_+ \uparrow G(m)$ of $G(m)$ induced from the representation $\Gamma_+$ of $G_0$ with respect to any coset decomposition of $G(m)$ with respect to $G_0$. From equation (2) it follows that every matrix corresponding to an arbitrary element of $G(m)$ commutes with all other matrices corresponding to other elements of it. Hence, $\Gamma_+ \uparrow G(m)$ is completely reducible, to one dimensional representations in which, as the order of each $\gamma_i$ is two the matrices corresponding to $\gamma_i$ are given by $\gamma_i \to \pm 1$. These representations are $2^m$ in number which is also the order of the factor group $G(m)/G_0$. Hence, these are the only possible representations of $G(m)$ in which $\bar{E} \to I = 1$.

*Class 2*: As there are $1 + 2^m(2 + 2^m)$ conjugate classes of $G(m)$ when $m$ is even (odd) there exists one (two) more nonequivalent irreducible representation $(S)$ of $G(m)$ of dimension greater than one. When $m = 2n$ (say) we designate it by $\Delta^-(2n)$ thereby indicating that in it $\bar{E} \to -I$. Now making use of the completeness relation, we find that the dimension of $\Delta^-(2n)$ is $2^n$.

When $m = 2n + 1$ (say), $\gamma_1 \gamma_2 \ldots \gamma_m = \delta_m$ commutes with all the elements of $G(m)$. Hence, the matrix corresponding to $\delta_m$ in an irreducible representation of $G(m)$ is either $kI$ or $-kI$ since $\delta_m^2 = \bar{E}^n$ and $k = i(1)$ if $n$ is odd or even. Hence, only two nonequivalent irreducible representations of dimension greater than one can exist and they should necessarily be of the same order. We designate them by $\Delta^-_\pm(2n + 1)$, thereby indicating that in them $\delta_m \to \pm kI$ and

$\bar{E} \rightarrow -I$. Now making use of the completeness relation, we find that the dimension of $\Delta_{\pm}^{-}(2n + 1)$ is $2^n$.

By the Mackey theory of induced representations $\Delta^{-}(2n)$ or $\Delta_{\pm}^{-}(2n + 1)$ should be obtained by reducing the induced representation $\Gamma_{-}(m) = \Gamma_{-} \uparrow G(m)$ whose dimension is same as the order of $G/G_0$ and is given by $2^m$. Now, obviously, $\Delta^{-}(2n)$ and $\Delta_{\pm}^{-}(2n + 1)$ are each contained $2^n$ times in $\Gamma_{-}(2n)$ and $\Gamma_{-}(2n + 1)$, respectively.

## 5. EXPLICIT FORMS OF $\Delta^{-}(2n)$ AND $\Delta_{\pm}^{-}(2n + 1)$

Consider the composition series

$$G(m) \supset G(m - 1) \cdots \supset G_0$$

terminating in $G_0$. In general, it is difficult to reduce $\Gamma_{-}(m)$ directly. Hence, we apply Mackey's method in steps through the composition series and obtain explicit forms of $\Delta^{-}(2n)$ and $\Delta_{\pm}^{-}(2n + 1)$.

Now the orbits containing the representations $\Delta^{-}(i)(\Delta_{\pm}^{-}(i))$ of $G(i)$, when $i$ is even (odd) relative to $G(i + 1)$, are given by $\{\Delta^{-}(i)\}$, $(\{\Delta_{+}^{-}(i), \Delta_{-}^{-}(i)\})$. For when $i$ is even, there exists only one nonequivalent irreducible representation of $G(i)$ and $\gamma_{i+1}$ should transform $\Delta^{-}(i)$ into itself as $\gamma_{i+1}$ commute with $\bar{E}$. When $i$ is odd $\gamma_{i+1}\delta_i\gamma_{i+1} = \bar{E}\delta_i$ and $\delta_i \rightarrow \pm kI$ in $\Delta_{\pm}^{-}(i)$, the two representations $\Delta_{\pm}^{-}(i)$ are transformed into each other by $\gamma_{i+1}$.

When $i$ is odd, the dimension of the induced representation $\Delta_{+}^{-}(i) \uparrow G(i + 1)$ is $2 \times 2^{(i-1)/2} = 2^{(i+1)/2}$, which is also the dimension of $\Delta^{-}(i + 1)$. Further as $\gamma_{i+1}$ commutes with $\bar{E}$, in the induced representation $\Delta_{+}^{-}(i) \uparrow G(i + 1)$ the $E \rightarrow -I \otimes I_2$. Hence, $\Delta_{+}^{-}(i) \uparrow G(i + 1)$ is irreducible and must be equivalent to the representation $\Delta^{-}(i + 1)$. Now making use of the coset decomposition $G(i + 1) = G(i) + \gamma_{i+1}G(i)$, the induced representation which is now identified with $\Delta^{-}(i + 1)$ is given by

$$\Delta^{-}(i + 1): \underline{\gamma}_{\gamma} \rightarrow \mathscr{L}_{\gamma} \otimes \sigma_1 \qquad \gamma = 1, \ldots, i$$
$$\gamma_{i+1} \rightarrow I \otimes \sigma_2 \tag{6}$$

where $\sigma_1 = \begin{pmatrix} 1 & 0 \\ 0 & -1 \end{pmatrix}$ and $\sigma_2 = \begin{pmatrix} 0 & 1 \\ 1 & 0 \end{pmatrix}$, and $I$ is a unit matrix of dimension $2^{(i-1)/2}$. Note that the above representations $\Delta^{-}(i + 1)$ are specified by giving the matrix representations of the generating elements of $G(i + 1)$. When $i$ is even from the above discussion, without loss of generality, $\Delta^-(i)$ can be taken as

$$\Delta^-(i): \quad \gamma_\gamma \to \mathscr{L}_\gamma \otimes \sigma_1 \qquad \gamma = 1, \ldots, i-1$$
$$\underline{\gamma_i} \uparrow I \otimes \sigma_2 \tag{7}$$

when $i \geq 2$. Since the $\gamma$'s anticommute with each other, the transform of $\Delta^-(i)$ by $\gamma_{i+1}$ is given by

$$\gamma_{i+1}\Delta^-(i)\gamma_{i+1}: \quad \gamma_\gamma \to \mathscr{L}_\gamma \otimes (-\sigma_1)$$
$$\underline{\gamma_i} \to I \otimes (-\sigma_2) \tag{8}$$

As $\gamma_{i+1}$ commutes with $\bar{E}$ and there exists only one nonequivalent irreducible representation of $G(i)$, with $\bar{E} \to -I$, $\gamma_{i+1}\Delta^-(i)\gamma_{i+1}$ must be equivalent to $\Delta^-(i)$. Hence, by Schur's lemma there exists a non-singular matrix $S$ such that

$$S\Delta^-(i) = \gamma_{i+1}\Delta^-(i)\underline{\gamma_{i+1}}S$$

Obviously, $S$ should be of the form $I \otimes \sigma_3$ and $\sigma_3$ should anticommute with $\sigma_1$ and $\sigma_2$. Hence, $\sigma_3$ is given by $\pm\begin{pmatrix} 0 & -i \\ i & 0 \end{pmatrix}$. Now the two nonequivalent representations of $G(i+1)$

$$\begin{aligned} \gamma_\gamma &\to \mathscr{L}_\gamma \otimes \sigma_1 \\ \gamma_i &\to I \otimes \sigma_2 \\ \gamma_{i+1} &\to \pm I \otimes \sigma_3 \end{aligned} \tag{9}$$

are of dimension $2^{i/2}$ and in them $\bar{E} \to -I \otimes I_2$. As $G(i+1)$ when $i$ is even has only two nonequivalent irreducible representations $\Delta^-_\pm(i+1)$ of dimension $2^{i/2}$ they must be equivalent to the above and, hence, we identify $\Delta^-_\pm(i+1)$ with them. To complete induction we note that when $i = 0$ with relavant representation $G_0$ is given by $\bar{E} \to -1$ and when $i = 1$ the representations $\Delta^-_\pm(1)$ of $G_1$ are given by $\bar{E} \to -1$, $\gamma_i \to \pm 1$. Note that only the representation $\Delta^-_\pm(1): \gamma_i \to 1$ need be considered in setting up the representations for different $G(m)$'s.

Now, obviously, from (6) and (9) the representation $\Delta^-(2n)$, $[\Delta^-_\pm(2n+1)]$ are given by

$$\gamma_1 \to \sigma_1 \otimes \sigma_1 \otimes \cdots \otimes \sigma_1 = \mathscr{L}_1$$
$$\gamma_{2j} \to \underbrace{I \otimes \cdots \otimes I}_{j \text{ times}} \otimes \sigma_2 \otimes \sigma_1 \otimes \cdots \otimes \sigma_1 = \mathscr{L}_{2j}$$
$$\gamma_{2j+1} \to I \otimes \cdots \otimes I \otimes \sigma_3 \otimes \sigma_1 \otimes \cdots \otimes \sigma_1 = \mathscr{L}_{2j+1}$$

with $n$ terms in each product. When $m$ is odd $\gamma_m$ should be taken with $\pm$ signs to obtain $\Delta_\pm^-(2n+1)$.


## 6. THE $\sigma$-OPERATION

Now we recover the $\sigma$-operation and find it identical with the above induction procedure. For that we write

$$L_m = \sum_{i=1}^{m} \mathscr{L}_i \lambda_i$$

when $m$ is even, and

$$L_m^\pm = \sum_{i=1}^{m-1} \mathscr{L}_i \lambda_i \pm \mathscr{L}_m \lambda_m$$

when $m$ is odd.

Now when $i$ is odd, multiplying $\mathscr{L}_\gamma$'s in (6) by $\lambda_\gamma$ adding, we obtain

$$L_{i+1} = \sum_{\gamma=1}^{i} \lambda_\gamma \mathscr{L}_\gamma \otimes \sigma_1 + \lambda_{i+1} I \otimes \sigma_2$$

$$= L_i^+ \otimes \sigma_1 + \lambda_{i+1} I \otimes \sigma_2$$

Since $i+1$ is even from (9) adding $\pm \lambda_{i+2} I \otimes \sigma_3$ to be above we obtain

$$L_{i+2}^\pm = L_i^+ \otimes \sigma_1 + I \otimes (\lambda_{i+1}\sigma_2 \pm \lambda_{i+2}\sigma_3)$$

This is the abstract form for the $\sigma$-operation. Obviously, the form of $L_{i+2}^\pm$ depends on the matrices used for the $\sigma$'s which anticommute with each other and the square of each is identity. For example, if we use the matrix representation

$$\sigma_1 \rightarrow \begin{pmatrix} 1 & 0 \\ 0 & -1 \end{pmatrix} \qquad \sigma_2 \rightarrow \begin{pmatrix} 0 & 1 \\ 1 & 0 \end{pmatrix} \qquad \sigma_3 \rightarrow \begin{pmatrix} 0 & -i \\ i & 0 \end{pmatrix} \qquad (11)$$

used in section 4 we obtain

$$L_{i+2}^\pm = \begin{bmatrix} L_i^+ & (\lambda_{i+1} \mp i\lambda_{i+2})I \\ (\lambda_{i+1} \pm i\lambda_{i+2})I & -L_i^+ \end{bmatrix}$$

But if we use

$$\sigma_1 \rightarrow \begin{pmatrix} 0 & 1 \\ 1 & 0 \end{pmatrix} \qquad \sigma_2 \rightarrow \begin{pmatrix} 0 & -i \\ i & 0 \end{pmatrix} \qquad \sigma_3 \rightarrow \begin{pmatrix} 1 & 0 \\ 0 & -1 \end{pmatrix}$$

obtained from (11) by a simple permutation of $\sigma$'s we obtain taking

$i = 2n - 1$

$$L_{2n+1}^{\pm} = \begin{bmatrix} \pm\lambda_{2n+1}I & L_{2n-1}^{+} - i\lambda_{2n}I \\ L_{2n-1}^{+} + i\lambda_{2n}I & \mp\lambda_{2n+1}I \end{bmatrix}$$

the $\sigma$-operation used in Ref. 1.

## REFERENCES

1. A. Ramakrishnan, "The Dirac Hamiltonian as member of a hierarchy of matrices," *J. Math Anal. and Appl.* **20**: 9–16 (1967).
2. H. Boerner, "Representations of Groups," North-Holland Publishing Co., Amsterdam, 1963.
3. P. Jordan and E. P. Wigner, *Z. Physik.* **47**: 631 (1928).
4. G. M. Mackey, I. *Ann. Math.* **55**: 101 (1952); II. *Ann. Math.* **58**: 193 (1952).
5. W. Pauli, "Handbuch der Physik 2nd Ed.," Berlin, J. Springer Verlag, V1, 1933.
6. W. Pauli, *Ann. Inst. Henri Poincaré*, **6**: 137 (1936).
7. J. S. Lomont, "Application of Finite Groups," Academic Press, New York, (1959).

# Relativistic Wave Equations for Higher Spin and Their Quantization

P. M. Mathews

*UNIVERSITY OF MADRAS*
*Madras, India*

---

## 1. INTRODUCTION

The scope of this paper will be limited to the presentation of certain recent results on relativistic wave equations which describe particles of definite (but arbitrary) spin $s$ and mass $m$, and on the quantization of these equations. To be more specific, the equations to be considered will have the Schrödinger form

$$i \frac{\partial \psi}{\partial t} (\mathbf{x}, t) = H \psi(\mathbf{x}, t) \tag{1}$$

with the Hamiltonian $H$ chosen in such a way as to leave (1) invariant under the transformations of the Poincaré group‡ as well as under the discrete transformations: $P$ (space inversion), $T$ (time reversal), and $C$ (charge conjugation). The derivation of such equations was the subject matter of my paper presented at the last Matscience symposium.[1] It is useful to recall here the main steps in this derivation before proceeding to the problem of second quantization. But let us start by enumerating the basic tenets of the philosophy behind equation (1).

---

‡We refer to the proper orthochronous homogeneous Lorentz group simply as the Lorentz group. The inhomogeneous group which includes space and time translations will be called the Poincaré group.

1. The equation must describe a particle of unique real mass $m$. This would be assured if

$$(p_0^2 - \mathbf{p}^2 - m^2)\,\psi = 0 \tag{2}$$

where

$$p_0 \equiv -i\frac{\partial}{\partial t} \quad \text{and} \quad \mathbf{p} \equiv -i\nabla \tag{3}$$

are the generators of space and time translations on the wave function. Equation (2) is nothing but the Klein–Gordon condition and will be satisfied, in view of (1), if

$$H^2 = E^2 \qquad E = +(p^2 + m^2)^{1/2} \qquad p = |\mathbf{p}| \tag{4}$$

This is a requirement on the Hamiltonian $H$ in (1). Lest this condition be taken too much for granted, let me remind the reader that though conventional relativistic wave equations[2] for higher-spin particles (with rare exceptions like Bhabha's multimass equation) do imply the Klein–Gordon condition (2), there has hardly ever been any attempt to cast the equations in the Schrödinger form (1), and even in cases where a Hamiltonian $H$ is available, it does not necessarily satisfy (4). A simple example is the Kemmer equation[3] for spin 1, which is reducible to a Schrödinger-type equation *plus* a supplementary condition: the Kemmer Hamiltonian does not obey (4), but rather, $H^3 = E^2 H$. Nevertheless, the Klein–Gordon equation is satisfied by virtue of the supplemetary condition.

2. The wave function $\psi$ must be locally covariant, i. e., the wave functions $\psi(x)$ and $\psi'(x')$, describing a given state as referred to two inertial reference frames, must have at any specified space-time point spinorial values which are related by a numerical matrix $\Lambda$ depending only on the Lorentz transformation between the two reference frames but not on the coordinates or on differential operators. Stated differently, the "spin part" of any Lorentz transformation operator on the wave function must be independent of coordinate and momentum variables. This property is taken for granted in all conventional equations, but it is certainly not indispensible, as is demonstrated by the formulation due to Foldy.[4] Indeed a change of representation through a suitable momentum-dependent unitary transformation will take a wave function which is locally covariant into another which is not.

The assumption of local covariance makes it possible to specify the transformation character of the wave function very simply by saying that it transforms according to an irreducible representation $D(m, n)$ of the Lorentz group or according to a direct sum of such representations, thus we may write $\psi \sim \sum \oplus \psi^{(m,n)}$.

2. The transformation of $\psi$ under the rotation subgroup of the Lorentz group must be such that the spin of the particle is uniquely fixed as $s$. This means that no repesentation of the rotation sub-group other than $D(s)$ should enter the picture; but this is true only of the representations $D(o, s)$ and $D(s, o)$ of the Lorentz group and none other, since the reduction of $D(m, n)$ with respect to the rotation sub-group is $D(m, n) \sim D(m + n) \oplus D(m + n - 1) \oplus \cdots \oplus D(|m - n|)$. Either $D(o, s)$ or $D(s, o)$ may be chosen if invariance under space inversion is not demanded of the theory, but if such invariance is required, $\psi$ must be taken to transform according to the $2(2s + 1)$-dimensional reducible representation $D(o, s) \oplus D(s, o)$ of the Lorentz group. This is what we assume. With the use of this type of wave function, it becomes impossible, for reasons outlined in Ref. 1, to obtain a manifestly covariant first order differential equation like the Dirac equation for any spin greater than $\frac{1}{2}$. Manifestly covariant equations of higher order do exist and, in fact, Weinberg[5] has obtained an equation of order $2s$, which however does not ensure that the Klein–Gordon condition (2) is satisfied. The latter has to be used as a supplementary condition on the wave function. On the other hand one could give up manifest covariance and look for an equation which is of the first order in the time derivative though not in the space derivatives, and which entails the Klein–Gordon equation automatically. It would be expected to contain, like the Dirac equation, more information than the Klein–Gordon equation taken by itself. This provides the motivation for the search for wave equtions of the form (1). But it must be remembered that it is the requirement that there be no redundant spins inherent in the wave function (which would later have to be eliminated by supplementary conditions) that is primarily responsible for our turning away completely from the conventional manifestly covariant forms of higher spin wave equations.

So far, the discussion was intended to motivate the investigation of relativistic equations of the form (1). We now outline the procedure for deriving such equations, and bring out the fact that there are four classes of equations which satisfy all the invariance conditions. The subsequent discussion will be aimed at proving that of all these equations, only two can be second-quantized in a manner consistent with reasonable criteria to be stated below; one of the two is applicable only to fermions of half-integral spin, and the other only to bosons of integral spin.

## 2. INVARIANCE CONDITIONS AND THE RELATIVISTIC HAMILTONIANS

The transformation character assumed above for the $2(2s + 1)$ component wave function $\psi$ determines the explicit forms of the generators $\mathbf{p}$ and $p_0$ of space and time translations, $\mathbf{J}$ of rotations and $\mathbf{K}$ of boosts on $\psi$, as follows:

$$p_0 = -\frac{i\partial}{\partial t} \tag{5a}$$

$$\mathbf{p} = -i\nabla \tag{5b}$$

$$\mathbf{J} = \mathbf{x} \times \mathbf{p} + \mathbf{S} \qquad \mathbf{S} = \begin{pmatrix} \mathbf{s} & 0 \\ 0 & \mathbf{s} \end{pmatrix} \tag{5c}$$

$$\mathbf{K} = t\mathbf{p} + \mathbf{x}p_0 + i\boldsymbol{\lambda} \qquad \boldsymbol{\lambda} = \begin{pmatrix} \mathbf{s} & 0 \\ 0 & -\mathbf{s} \end{pmatrix} = \rho_3\mathbf{S} \tag{5d}$$

where $(s_1, s_2, s_3) \equiv \mathbf{s}$ are the spin-$s$ angular momentum matrices and

$$\rho_1 = \begin{pmatrix} 0 & 1 \\ 1 & 0 \end{pmatrix} \qquad \rho_2 = \begin{pmatrix} 0 & -i \\ i & 0 \end{pmatrix} \qquad \rho_3 = \begin{pmatrix} 1 & 0 \\ 0 & -1 \end{pmatrix} \tag{6}$$

are the Pauli matrices whose elements are to be considered as multiples of the $(2s + 1)$-dimensional unit matrix. Further, the effect of space inversion $P$, time reversal $T$ and charge conjugation $C$ on $\psi$ are defined by

$$P\psi(\mathbf{x}, t) = \sigma\psi(-\mathbf{x}, t) \tag{7a}$$

$$T\psi(\mathbf{x}, t) = \tau\psi^*(\mathbf{x}, -t) \tag{7b}$$

$$C\psi(\mathbf{x}, t) = \kappa\psi^*(\mathbf{x}, t) \tag{7c}$$

The matrix $\sigma$ in (7a) can be chosen without loss of generality as $\rho_1$. The matrices $\tau$ and $\kappa$, however, are not unambiguously determined by the intrinsic properties of the operations $T$, $C$, and $P$ (essentially because commutation relations among these operators need be obeyed by their representatives only to within certain arbitrary phase factors. See, for instance, Ref. 4). In fact, even after taking into account their effects on the generators (5)—e.g., that time inversion should change the signs of $\mathbf{p}$ and $\mathbf{J}$ but should leave $p_0$ and

K unchanged—there is still a twofold uncertainty in $\tau$ and a similar uncertainty[6] in $\kappa$:

$$\tau = \begin{pmatrix} \zeta_s & 0 \\ 0 & e^{i\theta_t}\zeta_s \end{pmatrix} \qquad e^{i\theta_t} = \pm 1 \qquad TP = e^{i\theta_t}PT \qquad (8)$$

and

$$\kappa = \begin{pmatrix} 0 & \zeta_s \\ e^{i\theta_c}\zeta_s & 0 \end{pmatrix} \qquad e^{i\theta_c} = \pm 1 \qquad CP = e^{i\theta_c}PC \qquad (9)$$

where $\zeta_s$ is defined by

$$\zeta_s \mathbf{s}^* = -\mathbf{s}\zeta_s \qquad (10)$$

It is a direct consequence of this situation that the requirement of invariance of equation (1) under $T, C$, and $P$ and under the Poincaré group does not determine $H$ unambiguously. The invariance requirement is of course that the operator $H$ should be such as to behave exactly like $i\partial/\partial t \equiv -p_0$ under all these transformations. Specifically,

$$[H, p_0] = 0 \qquad [H, \mathbf{p}] = 0 \qquad [H, \mathbf{J}] = 0 \qquad [H, \mathbf{K}] = i\mathbf{p} \qquad (11)$$

and

$$PH = HP \qquad TH = HT \qquad CH = -HC \qquad (12)$$

When the conditions (11), (12), and (4) are applied to $H$ we get four possible classes of Hamiltonians corresponding to the four choices allowed by (8) and (9). There are four classes rather than just four Hamiltonians because the last of equations (11) is equivalent to a set of first-order ordinary differential equations whose solutions contain undetermined constants of integration.‡ The explicit expressions for the Hamiltonians are conveniently expressed in terms of projection operators $\Lambda_\nu$ to the eigenvalues $\nu$ of $\lambda_p \equiv (\lambda \cdot \mathbf{p}/p)$. The eigenvalues are $\nu = -s, -s+1, \ldots, s$ each of them occuring twice. If, for $\nu \geq 0$ we define

$$B_\nu = \Lambda_\nu + \Lambda_{-\nu} \qquad C_\nu = \Lambda_\nu - \Lambda_{-\nu} \qquad (13a)$$

then

$$B_\mu B_\nu = C_\mu C_\nu = B_\mu \delta_{\mu\nu} \qquad B_\mu C_\nu = C_\mu \delta_{\mu\nu} \qquad (13b)$$

---

‡The Hamiltonian obtained explicitly in Ref. 6 is a member of the class corresponding to the choice $TP = PT$, $CP = -PC$, and is the only one which goes to a proper limit as the momentum of the particle is made to vanish (this property being ensured by appropriate choice of the constants of integration).

and

$$\sum B_\nu = 1 \tag{13c}$$

In terms of these, the various forms possible for the Hamiltonian depending on the choice of signs made in (8) and (9), can be written as follows:‡

*Case (i)*: $TP = PT, CP = -PC$.

$$H = E \sum \tanh (2\nu\theta + \eta_\nu) C_\nu + \rho_1 E \sum \text{sech} (2\nu\theta + \eta_\nu) B_\nu \tag{14}$$

*Case (ii)*: $TP = -PT, CP = PC$.

$$H = E \sum \coth (2\nu\theta + \eta_\nu) C_\nu + i\rho_1 E \sum \text{cosech} (2\nu\theta + \eta_\nu) B_\nu \tag{15}$$

*Case(iii)*: $TP = PT, CP = PC$.

$$H = E \sum \coth (2\nu\theta + \eta_\nu) C_\nu + \rho_1 E \sum \text{cosech} (2\nu\theta + \eta_\nu) C_\nu \tag{16}$$

*Case(iv)*: $TP = -PT, CP = -PC$.

$$H = E \sum \tanh (2\nu\theta + \eta_\nu) C_\nu + i\rho_1 E \sum \text{sech} (2\nu\theta + \eta_\nu) C_\nu \tag{17}$$

Here $\theta$ is defined by

$$\sinh \theta = \frac{p}{m} \quad \text{and} \quad \cosh \theta = \frac{E}{m} \tag{18}$$

and the $\eta_\nu$ are constants of integration which remain entirely arbitrary. However, it will now be shown that if second quantization of equation (1) is to be carried out, only the Hamiltonians of cases (i) and (iii), i.e., equations (14) and (16), will be admissible, and even in these, all the $\eta_\nu$ must be taken to be zero. This severe restriction comes about as a consequence of the following microcausality condition§ imposed on $\psi$, considered now as a quantized field:**

$$[\psi_\alpha(\mathbf{x}, t), \psi_\beta^*(\mathbf{y}, \tau)]_\pm = 0 \quad \text{if } |\mathbf{x} - \mathbf{y}|^2 - (t - \tau)^2 > 0 \tag{19}$$

---

‡For details of derivation, see Ref. 6 and P. M. Mathews and S. Ramakrishnan, *Nuovo Cim.*, **50**: 339 (1967).

§Equation (19) will imply microcausality in the sense of commutability of observable densities at points separated by spacelike distances if such densities are local functions of the field $\psi$. That this is indeed the case for momentum density, energy density etc., in the present theory will be observed later.

**We use asterisks to denote Hermitian conjugation of Hilbert space operators and complex conjugation of c-numbers. A dagger will imply, in addition, a transposition in the finite dimensional spinor space.

Proof of the above statement is given most directly by evaluating the expression on the left-hand side of (19) with the aid of fermion or boson commutation rules applied to operators which appear as coefficients in the expansion of $\psi$ and $\psi^\dagger$ in terms of plane wave solutions of the Schrödinger equation (1). It is useful to present some properties of these solutions before embarking on the proof. Incidentally, it may be mentioned that, curiously enough, the question whether the plus sign (anticommutator) or minus sign (commutator) in (19) is to be taken is itself decided by the microcausality requirement.

Plane wave solutions of equation (1) when any one of the Hamiltonians (14)–(17) is used can be labeled by the complete set of quantum numbers $\mathbf{q}$, $\epsilon$ and $h$ which are eigenvalues of the commuting set of observables $\mathbf{p}$ (momentum), $H/E$ (sign of the energy), and $\mathbf{s}\cdot\mathbf{p}/p$ (helicity). Such solutions possess the orthonormality property

$$\int \psi_{\mathbf{q}h}^{\epsilon\dagger}(\mathbf{x}, t)\, M\, \psi_{\mathbf{q}'h'}^{\epsilon'}(\mathbf{x}, t)\, d^3x = \eta\delta(\mathbf{q} - \mathbf{q}')\,\delta_{\epsilon\epsilon'}\delta_{hh'} \qquad (20)$$

In the above, $M \equiv M(\mathbf{p})$ is a matrix-differential operator, the presence of which is required in order to make the integral Lorentz-invariant and thus ensure that the orthonormality is an invariant statement; it plays the role of a metric operator in terms of which the Lorentz invariant scalar product $(\phi, \psi)$ between any two solutions $\phi, \psi$ of (1) is defined:

$$(\phi, \psi) = \int \phi^\dagger(\mathbf{x}, t)\, M\, \psi(\mathbf{x}, t)\, d^3x \qquad (21)$$

There are two possibilities for the metric operator, a positive definite one which we denote by $M_1$, and an indefinite one $M_2 \equiv M_1(H/E)$. And the constant $\eta$ to which normalization is done in (20) is dependent on this choice:

$$\eta = 1 \qquad \text{if } M = M_1 \qquad (22a)$$

and

$$\eta = \epsilon \qquad \text{if } M = M_2 \qquad (22b)$$

Explicit expressions for $M_1$ and $M_2$, obtainable by requiring the invariance of (21) under space inversion and under the Poincaré group, will be given later. They depend on the particular Hamiltonian chosen. We shall now proceed with the general treatment of the quantization process, which does not require knowledge of the actual forms of $H$ and $M$ till the very last step.

## 3. MICROCAUSALITY AND QUANTIZATION

The primary criterion on which our quantization procedure is based is the microcausality condition (19). To understand its implications, we evaluate the left-hand side by writing

$$\psi(\mathbf{x}, t) = \int d^3q \left\{ \sum_h a(\mathbf{q}, h) \, \psi_{qh}^+(\mathbf{x}, t) + \sum_h b^*(\mathbf{q}, h) \, \psi_{-qh}^-(\mathbf{x}, t) \right\} \quad (23)$$

and using the commutation rules

$$[a(\mathbf{q}, h), a^*(\mathbf{q}', h')]_\pm = [b(\mathbf{q}, h), b^*(\mathbf{q}', h')]_\pm = \delta(\mathbf{q} - \mathbf{q}') \delta_{hh'} \quad (24)$$

all other anticommutators (commutators) being zero. It is a simple matter to reduce the left-hand side of (19), using (23) and (24), to

$$[\psi_\alpha(\mathbf{x}, t), \psi_\beta^*(\mathbf{y}, \tau)]_\pm = S_{\alpha\beta}^+ \pm S_{\alpha\beta}^- \quad (25)$$

where

$$S^\epsilon(\mathbf{x} - \mathbf{y}, t - \tau) = \int d^3q \sum_h \psi_{qh,\alpha}^\epsilon(\mathbf{x}, t) \psi_{qh,\beta}^{\epsilon*}(\mathbf{y}, \tau) \quad (26)$$

The sums over positive and negative energy solutions of (1), involved in the above expression, must evidently depend on the Hamiltonian. In fact, it can be shown[7] that

$$S^\epsilon = (2\pi)^{-3} \int d^3q \, \eta M^{-1}(\mathbf{q}) \frac{1}{2} \left[ 1 + \frac{H(\mathbf{q})}{\epsilon\omega} \right] \exp\left[ i\mathbf{q} \cdot (\mathbf{x} - \mathbf{y}) - i\epsilon\omega(t - \tau) \right] \quad (27)$$

It will be noticed that $H$ and $M$ in (27) contain the numerical momentum vector $\mathbf{q}$ in the place of the differential operator $\mathbf{p}$. It is immaterial whether $\eta M^{-1}(\mathbf{q})$ is taken as $M_1^{-1}(\mathbf{q})$ or $\epsilon M_2^{-1}(\mathbf{q}) \equiv \epsilon M_1^{-1}(\mathbf{q}) H(\mathbf{q})/\omega$, since both these forms are identical when operating on the remaining factors in the integrand which constitute an eigenfunction of $H(\mathbf{q})$ with eigenvalue $\epsilon\omega$. We will use either of the alternative forms, depending on convenience.

The remaining task is to determine, for each of the Hamiltonians (14)–(17), whether it is $(S^+ + S^-)$ or $(S^+ - S^-)$ which vanishes for $|\mathbf{x} - \mathbf{y}|^2 - (t - \tau)^2 > 0$, or whether neither one vanishes. It turns out that with the Hamiltonian of case (i), equation (14), only $(S^+ + S^-)$ can be causal and only under the condition that all the $\eta_\nu$ be zero and that the spin be half-integral; thus it is suitable only for the description of fermions of half-integral spin. The Hamiltonian of case (iii), equation (16), leads to $(S^+ - S^-)$ being causal, again with the proviso that all the $\eta_\nu$ be zero, but the spin must be integral; this Hamiltonian is suitable therefore only for bosons of integral spins. The remaining two Hamiltonians do not allow either $(S^+ + S^-)$ or $(S^+ - S^-)$ to be causal, and therefore have to be considered as unphysical in the context of a quantized field theory.

Proof of the above results rests on the observation that is intimately related to the invariant functions $\Delta^\epsilon$ which are solutions of the Klein–Gordon equation and are defined by

$$\Delta^\epsilon(\mathbf{x}, t) = \frac{-i\epsilon}{(2\pi)^3} \int \frac{d^3q}{2\omega} \exp\left[i(\mathbf{q} \cdot \mathbf{x} - \epsilon\omega t)\right] \tag{28}$$

These functions have the property that $\Delta \equiv \Delta^+ + \Delta^-$ is causal:

$$\Delta(\mathbf{x}, t) \equiv \Delta^+(\mathbf{x}, t) + \Delta^-(\mathbf{x}, t) = 0 \text{ if } \mathbf{x}^2 - t^2 > 0 \tag{29}$$

while no other combination of $\Delta^+$ and $\Delta^-$ is causal. Now, if the factor

$$i\eta M^{-1}(\mathbf{q})\{\epsilon\omega + H(\mathbf{q})\} \tag{30}$$

is pulled out of the integral (27) by replacing $\mathbf{q}$ by the differential operator $\mathbf{p}$ and $\epsilon\omega$ by $i\partial/\partial t$, then $S^\epsilon$ reduces to the form of a matrix differential operator $F$ acting on $\Delta^\epsilon$. If $F$ happens to be independent of $\epsilon$ then $(S^+ + S^-)$ becomes $F(\Delta^+ + \Delta^-) = F\Delta$, which would be causal provided $F$ is a local operator. On the other hand, if $F = \epsilon F'$ where $F'$ is independent of $\epsilon$, then it is $(S^+ - S^-)$ that would be causal. If $F$ has any other kind of dependence on $\epsilon$, neither $(S^+ + S^-)$ nor $(S^+ - S^-)$ can be causal. With these considerations in mind, let us now take a closer look at the case when the Hamiltonian of equation (14), together with the associated metric operator, is introduced in (27). This would serve as an illustration of the manner in which the specific results stated above for the four different cases are obtained.

The metric operator $M_1$ associated with $H$ of case (i), equation

(14) [as also with the $H$ of case (iv), equation (17)], is‡

$$M_1 = \left(\frac{E}{m}\right) \sum \text{sech } (2\nu\theta + \eta_\nu) B_\nu \qquad (31)$$

Thus, the expression (30) in this case becomes

$$im \left[\sum \epsilon \cosh (2\nu\theta + \eta_\nu) B_\nu + \sum \sinh (2\nu\theta + \eta_\nu) C_\nu + \rho_1\right] \qquad (32)$$

In (32), $\theta$, $B_\nu$, and $C_\nu$ are to be taken as being defined with the numerical quantities $\mathbf{q}$ and $\omega$ in the place of the operators $\mathbf{p}$ and $E$. Observe now that since (32) contains one term, $\rho_1$, which is clearly independent of $\epsilon$, the only way of ensuring causality would be by having the same property for the other terms also, so that the whole expression would go over into an $F$ independent of $\epsilon$. This in turn demands that cosh $(2\nu\theta + \eta_\nu)$ must be an odd function of $\omega$, which would enable us to couple one factor of $\omega$ with $\epsilon$ and then to replace $\epsilon\omega$ by $i\partial/\partial t$, removing the explicit $\epsilon$-dependence. Similarly, sinh $(2\nu\theta + \eta_\nu)$ should be an even function of $\omega$, since an odd factor of $\omega$ would be equivalent to $\epsilon\, i\partial/\partial t$. Both these are achieved if (and only if) $\eta_\nu = 0$ and $2\nu$ is odd (which implies that the spin is half-integral). This may be seen from the $\omega$-dependence of sinh $2\nu\theta$ and cosh $2\nu\theta$:

$$\sinh 2\nu\theta = \sum_{n \text{ odd}} \alpha_n \sinh^n \theta = \sum_{n \text{ odd}} \alpha_n \left(\frac{q}{m}\right)^n \qquad (33a)$$

and

$$\cosh 2\nu\theta = \sum_{n \text{ odd}} \beta_n \cosh^n \theta = \sum_{n \text{ odd}} \beta_n \left(\frac{\omega}{m}\right)^n \qquad (33b)$$

if $2\nu$ is odd, the $\alpha_n$ and $\beta_n$ being constants. (If $2\nu$ is even it is sinh $2\nu\theta$ which has odd powers of $\omega$ and cosh $2\nu\theta$ contains just even powers of $\omega$.) Noting that $\sum \cosh 2\nu\theta \cdot B_\nu = \cosh 2\lambda_p \theta$ and $\sum \sinh 2\nu\theta\, C_\nu = \sinh 2\lambda_p \theta$, we conclude that under the conditions stated above, $S^\epsilon$ reduces to

$$S^\epsilon = F\,\Delta^\epsilon \equiv im \left[E^{-1} \cosh 2\lambda_p\theta\, i\frac{\partial}{\partial t} + \sinh 2\lambda_p\theta + \rho_1\right]\Delta^\epsilon \qquad (34)$$

Since $E^{-1} \cosh 2\lambda_p\theta$ and $\sinh 2\lambda_p\theta$ are multinomials§ in the com-

---

‡The special case of (31) for $\eta_\nu = 0$ is derived in P. M. Mathews, *Phys. Rev.* **143**: 985 (1966). The same method may be used in the more general cases too. Explicit expressions for $M_1$ and $M_2$ corresponding to all the Hamiltonians (14)–(17) are given in P. M. Mathews and S. Ramakrishnan, *Nuovo Cim.* **50**: 339 (1967).
§See the appendix to Ref. 5

ponents of $\mathbf{p}$, $F$ is a local operator and it follows that $S^+ + S^- = F\Delta$ is a causal function, implying that the anticommutator in (25) is causal.

To sum up, the Schrödinger equation (1) with the Hamiltonian (14) can be quantized consistently with microcausality only if all the $\eta_\nu = 0$, the spin is half-integral and fermion rules are employed. A similar analysis can be carried out with the Hamiltonians (15)–(17) and the associated metrics, namely, (31) in case (iv) and

$$M_2 = \left(\frac{E}{m}\right) \sum \operatorname{cosech} (2\nu\theta + \eta_\nu) C_\nu \qquad (35)$$

in cases (ii) and (iii). The results are as follows:

With the Hamiltonian of case (iii), equation (16), quantization is successful provided $\eta_\nu = 0$, the spin is integral and boson commutation rules are employed. In the remaining cases (ii) and (iv), though an expression of the form $F\Delta$ can be obtained for the commutator or anticommutator in (25), $F$ is not a local operator so that causality is not ensured.

Perhaps the most remarkable aspect of the above treatment is the power of the microcausality condition: to reject Hamiltonians as unacceptable or, when acceptable, to determine the values of the parameters $\eta_\nu$, to fix the nature of the spin (integral or half integral) and, finally, to assign the statistics (boson or fermion)—the assignment being fortunately in agreement with the conventional spin-statistics connection! However, there is one thing it does not determine, namely, whether the definite metric $M_1$ or the indefinite one $M_2$ should be used in any given case. As is evident from the remarks below equation (27), the choice of metric is irrelevant as far as microcausality is concerned. But the metric is important in connection with the role of the total energy and momentum operators of the field (defined as $\int \psi^\dagger MH\psi\, d^3x$ and $\int \psi^\dagger M\mathbf{p}\psi\, d^3x$) as translation generators on the field. It turns out that $M_1$ must be chosen for fermion fields and $M_2$ for boson fields.

One final remark on the appropriateness of (19) as the microcausality condition, which as observed (see footnote p. 38) depends on observable densities being local functions of the fields: It is easily verified that $M_1$ in case (i) (half-integral spin fermion field) and $M_2$ in case (iii) (integral spin boson field) are both of the form $(2m)^{-1}(\rho_1 H + H^+ \rho_1)$ so that the total charge, $\int \psi^\dagger M\psi\, d^3x$ can be rewritten with the aid of equation (1) as

$$\frac{1}{2m} \int \left[ \psi^+ \rho_1 i \frac{\partial \psi}{\partial t} - i \frac{\partial \psi^+}{\partial t} \rho_1 \psi \right] d^3 x \qquad (36)$$

The integrand, which may be taken as the charge density, is clearly a local function of the field. The integrals for the total energy operator etc., can be rewritten in a similar form in which the locality of the integrands as functions of $\psi$ is obvious.

## 4. CONCLUSION

The theory we have presented is unconventional in that it is based on a wave equation which is not manifestly covariant and makes no use of any Lagrangian. The first derivation of a relativistic Schrödinger equation for arbitrary spin was by Weaver, Hammer, and Good,[8] who started from an assumed rest-system Hamiltonian and obtained essentially our Hamiltonian (14) with $\eta_v = 0$ but did not realize the existence of the other three possibilities. The work of Weinberg[5] is another example of a theory involving wave functions transforming according to the representation $D(o, s) \oplus D(s, o)$. Our wave functions for plane waves in the two quantizable cases coincide with those obtained by Weinberg using Wigner's[9] little-group approach. In the integral spin case, there is the peculiarity that the positive and negative energy wave functions belonging to the same momentum and zero helicity are identical. This is reflected in the nature of the relevant Hamiltonian

$$H = E (\coth 2\lambda_p \theta + \rho_1 \operatorname{cosech} 2\lambda_p \theta) \qquad (37)$$

[obtained by setting $\eta_v = 0$ in (16)], which becomes singular as $\theta \to 0$ (limit of vanishing momentum) in the sense that the elements of $H$ become infinite in this limit but in such a way that the eigenvalues tend to $\pm m$. It is to be particularly noted that these peculiarities result not from the assumption of the Schrödinger form for the wave equation, but from the transformation character of the wave function itself, assumed to be according to $D(o, s) \oplus D(s, o)$. There remains the question whether it is possible to introduce interactions consistently into the theory; in this respect the lack of manifest covariance is a disadvantage, while the absence of subsidiary conditions which are the main stumbling blocks in the case of conventional theories is a distinct advantage. The answer to the question must however await the results of further investigation.

## REFERENCES

1. "Symposia on Theoretical Physics and Mathematics," Vol. 6, Ed. Alladi Ramakrishnan, Plenum Publishing Corp. New York, 1967.
2. E. M. Corson, "Tensors, Spinors and Relativistic Wave Equations," Blackie, London, 1954.
3. N. Kemmer, *Proc. Roy. Soc.* A **173**: 91 (1939).
4. L. L. Foldy, *Phys. Rev.* **102**: 568 (1956).
5. S. Weinberg, *Phys. Rev.* **133**: B1318 (1964).
6. P. M. Mathews, *Phys. Rev.* **143**: 978 (1966).
7. P. M. Mathews, *Phys. Rev.*, **155**: 1415 (1967).
8. D. L. Weaver, C. L. Hammer and R. H. Good, Jr., *Phys. Rev.* **135**: B241 (1964).
9. E. P. Wigner, *Ann. Math.* **40**: 149 (1939).

# Contributions to the Relativistic Generalization of the Kinetic Theory of Gases and Statistical Mechanics

J. I. HORVÁTH

*JÓZSEF ATTILA UNIVERSITY‡*
*Szeged, Hungary*

## 1. INTRODUCTION

The concept of the space–time world represents the geometrization of the space and time relations of the physical system considered. While in the framework of the special theory of relativity only the kinematic aspects of the space and time relations are characterized and the space–time continuum as an underlying geometrical background of physical events has been considered, in Einstein's theory of gravitation the space–time continuum has significantly other meaning, namely, its geometrical structure is determined by the gravitational interactions, therefore, the space–time continuum also represents a geometrization of the gravitational field.

Somewhat analogous geometrization of the kinematic and dynamic relations has been introduced in the framework of the phase-space formalism of the nonrelativistic gas theory suggested earlier by the kinetic and statistic theories of Boltzmann.

This means, however, that having a relativistic generalization of the phase-space formalism in mind, the concept of the relativistic phase-space has to reveal the geometrization—in other terms: a

---

‡Department of Theoretical Physics.

geometrical mapping—of the space–time *and* dynamic relations of the physical system considered.

If a relativistic phase-space formalism has to be developed, one must introduce in every point of the four-dimensional space–time continuum a *local momentum space* which is, however, only three-dimensional due to the familiar normalization condition of the four-momenta; hence, the relativistic phase space of the gaseous particles—or in terms of the terminology of the Ehrenfests: the relativistic $\mu$-space—is (4 + 3)-dimensional.

In fact, the points, i.e., the "radius vectors" of the momentum space are potentially arbitrary but, actually, as the tangents of the world-lines of the particles they are governed in all points of the space–time by the equations of motion, i.e., by the dynamical space and time relations of the systems considered. The potential arbitrariness of the momenta means that the momentum-space as a geometrical mapping of the internal degrees of freedom of system dynamics localized at every point of the space–time is determined unambiguously by the potentially arbitrary initial values of motion.

In terms of the phase-space formalism the trajectories of the gaseous particles reflect their dynamical history and depend on the initial values mentioned. As a matter of fact, the dynamical relations of the system at every point of the space–time continuum are determined by the tangents of the trajectories (world-lines), i.e., by the actual momentum of the particle considered. Therefore, the dynamical relations of the system depending on the dynamical history of its constituents can be geometrized only if a geometry could be introduced in which the geometrical and physical quantities at every point of the space considered are dependent on the directions, too. Such a geometrical framework means the geometry of the *general line-element space* where the geometrical quantities in every point $x \equiv \{x^\mu\}$ are dependent on the homogeneous direction coordinates $v \equiv \{v^\mu\}$ ($\mu = 0, 1, 2, 3$).[1] Some aspects of the relativistic phase-space formalism were discussed in a previous paper[2] in order to find any connection between the phase-space formalism and the internal structure of physical fields, i.e., in order to hint at a possible geometrization of isobaric spin space.

Now, we would like to point out that, in fact, beside the space and time relations, the dynamical relations of the system considered are geometrized in the framework of the line-element geometry by the dependence of the geometrical and physical quantities on the

direction. This means, however, a second step in the geometrization; therefore, this idea may be denoted as a kind of "hypergeometrization" based on a geometry of an adequate four-dimensional
anisotropic space.[3]

The geometric anisotropy of the space means that all of the
geometrical quantities, particularly the components of the metrical
foundamental tensor $g_{\mu\nu}$—the most important geometrical quantity
determining the geometrical structure of the line-element space—
depend not only on the space point $\{x^\mu\}$ but also on the homogeneous
direction coordinates $\{v^\nu\}$, i.e., $g_{\mu\nu} = g_{\mu\nu}(x^\mu, v^\mu)$ being a homogeneous
function of the variables $v^\mu$ of zero degree. The homogeneity of the
direction coordinates means, of course, that the $v^\mu$ are not independent of each other and only their proportion have meaning.

The transformation law of the line elements are given by

$$x^{\mu'} = x^{\mu'}(x^\mu), \ v^{\mu'} = v^\mu \frac{\partial x^{\mu'}}{\partial x^\mu} \qquad \left( \Delta \equiv \det \left| \frac{\partial x^{\mu'}}{\partial x^\mu} \right| \neq 0 \right) \qquad (1)$$

If the directions determined by the direction coordinates $\{v^\mu\}$ are
the same as those of the potentially arbitrary momenta, one can
immediately see that the line-element space could be the natural
geometrical background of the relativistic phase-space being also
$(4 + 3)$-dimensional as well. The group of transformations (1)—
which will be called in the following as the group $G_x$—does not
take all the advantages of the geometrical features of the line-
element geometry which may be used very favorably for our
purposes.

In order to emphasize the fact that only three of the homogeneous direction coordinates $\{v^\mu\}$ are independent, let us introduce
inhomogeneous direction coordinates as follows.

First of all based on the definition of the so-called metrical
fundamental function:

$$F(x, v) \equiv \{g_{\mu\nu}(x, v)v^\mu v^\nu\}^{1/2} \qquad (2)$$

which is a homogeneous function of the variables $v^\mu$ of first order,
one can introduce a vector of unit length having the same direction
as that of the line element $\{x^\mu, v^\mu\}$ determined by $\{v^\mu\}$

$$l^\nu \equiv \frac{v^\mu}{F} = \frac{p^\mu}{m_0} \qquad (3)$$

due to the above considerations and to the familiar equation of normalization of the four-momenta

$$p^\mu p_\mu = m_0^2 \tag{4}$$

as well. (The velocity of light is taken as unit velocity.)

Now, let us introduce in every point of the space–time continuum a so-called $\lambda$-*triad* formed by three mutually orthogonal vectors $\lambda^\mu_i$ ($i = 1, 2, 3$) of unit length being in the local rest frame of reference $\mathcal{K}^\circ$ of the gaseous particles orthogonal to the timelike four-momentum $p_0^\mu$ of the gaseous particles. Due to the metrical structure of the line-element space, the orthonormality of the $\lambda^\mu_i$ vectors and their orthogonality to $p_0^\mu$ can be formulated as

$$g_{\mu\nu}(x, v)\lambda^\mu_i \lambda^\nu_j = -\delta_{ij} \tag{5}$$

$$g_{(0)\mu\nu}(x, v)\lambda^\mu_i p_0^\nu = 0 \qquad (i, j = 1, 2, 3) \tag{6}$$

The three independent degrees of freedom of the momentum space may be characterized by the angles

$$\vartheta_i \equiv \text{arc cos} \left\{ \frac{g_{\mu\nu}(x, v)\lambda^\mu_i p^\nu}{m_0} \right\} \tag{7}$$

between the directions corresponding to the radius vectors of the momentum space and the unit vectors $\lambda^\mu_i$. The angles $\vartheta_i$ and more favorably the scalar quantities

$$\xi_i \equiv \frac{g_{\mu\nu}(x, v)\lambda^\mu_i p^\nu}{m_0} \tag{8}$$

respectively, which are uniquely determined up to the sign of $p^0$, can be defined as *inhomogeneous direction coordinates*.

Of course, the four unit vectors $p^\mu/m_0$ and $\lambda^\mu_i$, respectively, form an orthonormal tetrad with the axes $\Lambda^\mu_{(0)} \equiv p^\mu/m_0$ and $\Lambda^\mu_{(i)} \equiv \lambda^\mu_i$. Having the generalization of the tetrad-formalism of the Riemanian space–time world in mind,[4] if the covariant components of the tetrad-axes are defined by

$$\Lambda_{(\alpha)\mu} = g_{\mu\nu}\Lambda^\nu_{(\alpha)} \qquad (\alpha = 0, 1, 2, 3) \tag{9}$$

where the indices in parentheses like $(\alpha)$ mean a label distinguishing the particular axes, the condition of orthonormality of equations (4) and (5) may be written in the form

$$\Lambda^\mu_{(\alpha)}\Lambda_{(\beta)\mu} = \eta_{(\alpha\beta)} \tag{10}$$

where

$$\eta_{(00)} = -\eta_{(11)} = -\eta_{(22)} = -\eta_{(33)} = 1 \qquad \eta_{(\alpha\beta)} = 0 \ (\alpha \neq \beta)$$

$$\eta_{(\alpha\beta)} = \eta^{(\alpha\beta)} \tag{11}$$

is a diagonal matrix satisfying the relations

$$\eta^{(\alpha\beta)} \eta_{(\beta\gamma)} = \delta_\gamma^\alpha \tag{12}$$

being, in language of matrix algebra, a square of unity.

One has to emphasize that the labels on the vectors have no tensorial meaning; nevertheless, by means of the $\eta$-matrix the framework of the tensor calculus can be introduced. Let the raising and lowering the labels be defined by

$$\Lambda^{(\alpha)\mu} = \eta^{(\alpha\beta)} \Lambda_{(\beta)}^\mu, \ \Lambda_{(\alpha)\mu} = \eta_{(\alpha\beta)} \Lambda_\mu^{(\beta)}, \ \text{etc.,} \tag{13}$$

then the two tetrads $\Lambda_{(\alpha)}^\mu$ and $\Lambda^{(\alpha)\mu}$ are closely connected: their spacelike axes are the same and their timelike axes are opposite to one another, i.e., they are in their handedness different.

Let us give at a space–time point two orthonormal tetrads $\Lambda_{(\alpha)}^\mu$ and $\tilde{\Lambda}_{(\tilde{\alpha})}^\mu$, respectively, they can be connected by a Lorentz transformation with the *Lorentz matrix*

$$L_{.(\tilde{\alpha})}^{(\alpha)} \equiv \Lambda_\mu^{(\alpha)} \tilde{\Lambda}_{(\tilde{\alpha})}^\mu \tag{14}$$

being just the unit matrix if the two tetrads coincide. This means, however, that at every point of the space the equivalent Lorentz transformation

$$\tilde{\Lambda}_{(\tilde{\alpha})}^\mu = L_{.(\tilde{\alpha})}^{(\alpha)} \Lambda_{(\alpha)\mu} \qquad \text{and} \qquad \Lambda_\mu^{(\alpha)} = L_{.(\tilde{\beta})}^{(\alpha)} \tilde{\Lambda}_\mu^{(\beta)} \tag{15}$$

can be introduced, being independent of any changes of the space–time coordinates $\{x^\mu\}$. These Lorentz transformations may be interpreted as the *internal changes of the orientation* of the tetrads.

Tetrads at different points of the space–time world may be compared either by a Fermi–Walker transport of the tetrad axes along the world lines of the gaseous particles,[4] or in special, but from physical point of view enough general case of line-element spaces by the absolute parallelism of line elements.

Keeping the familiar definition of the Fermi coordinates

$$\Xi_{(\alpha)} \equiv \frac{g_{\mu\nu} \Lambda_{(\alpha)}^\mu p^\nu}{m_0} \tag{16}$$

of a unit vector $p^\mu/m_0$ in mind,[4] our inhomogeneous direction coordinates $\xi_i$ are, of course, the spacelike part of the $\Xi_{(\alpha)}$, i.e., $\xi_i = \Xi_{(i)}$.

Both the Fermi coordinates and the inhomogeneous direction coordinates are invariants of the general group $G_x$ of the coordinate transformations, but they depend on the orientation of the tetrad and $\lambda$-triad, respectively. As a matter of fact, the Lorentz transformations (15) as well as their three-dimensional subgroup defined in a three-dimensional spacelike hypersurface of the space–time world by the changes of the orientation of the $\lambda$-triads as *internal transformations of the physical systems* can be interpreted, which cannot be induced by any change of the coordinates $\{x^\mu\}$. Therefore, based on the proposed method of hyperquantization some internal dynamical feature of the gaseous system considered can be characterized in terms of the Lorentz group (15) or by the group $G_\xi$ of the $\lambda$-triad as well. These groups are internal in such a sense that they characterize rather dynamical relations, e.g., dynamical symmetries, of the physical systems than space–time relations.

Due to the dimensionality of the momentum-space it seems that in our case the three-dimensional rotary-reflection groups $G_\xi$ of the $\lambda$-triad is especially important. For example, if the considered gas system streams with constant velocity, i.e., a certain direction is distinguished in the momentum space, the phase space loses its isotropy and the metrical fundamental tensor will be explicitly dependent on the directions characterized by the momentum.

This means, however, that the general group of the relativistic phase-space is the direct product of the groups $G_x$ and $G_\xi$:

$$G = G_x \otimes G_\xi \tag{17}$$

## 2. DEFINITION OF THE RELATIVISTIC PHASE-SPACE VOLUME

In order to find out the definition of the relativistic phase-space volume we have to suggest an adequate definition of the hypersurfaces in the framework of the general line-element geometry.

As a hypersurface of the geometrized relativistic phase-space the ensemble of the line elements $\{x^\mu = x^\mu(u^i), p^\mu\}$ or in terms of the inhomogeneous direction coordinates $\{x^\mu = x^\mu(u^i), \xi_i\}$ $(i = 1, 2, 3)$ will be denoted where $u^i$ and $u^i = $ const. mean, respectively, the parameters and the parametric lines of the three-dimensional hypersurface of the coordinates of the line elements $\{x^\mu, p^\mu\}$.

Due to this definition of the hypersurface the invariant volume element of the coordinate space $\{x^\mu\}$ can be constructed as it is well known in the Riemannian geometry. In the particular important special case of the parametrization

$$x^0 \equiv s, \ x^i \equiv u^i \qquad (i = 1, 2, 3) \tag{18}$$

where $s$ is the parameter of the length of arc of the line-element geometry, and if it is supposed that the unit tangent of the curve $x^\mu = x^\mu(s)$ coincides in the crossing point of the curve and the hypersurface considered with the unit normal vector of the hypersurface, the formally well-known formula

$$dV \equiv \sqrt{-g} \, dx^0 dx^1 dx^2 dx^3 \equiv \sqrt{-g} \, d^4x \tag{19}$$

can be obtained.[5]

Owing to the obvious invariance of $d\xi_i$ against any coordinate transformation of the group $G_x$ the invariant volume element of the local momentum space at arbitrary but fixed point of the coordinate space can, of course, be defined as follows:

$$dP \equiv m_0^3 d\xi_1 d\xi_2 d\xi_3 \equiv m_0^3 d^3\xi \tag{20}$$

where the factor $m_0^3$ has to be introduced in order to save the correct physical dimensions of the volume element of the momentum space.[5]

Although $dP$ is an invariant of the group $G_x$, it will generally change if the internal transformations

$$\xi_{i'} = \xi_{i'}(\xi_i) \tag{21}$$

of the group $G_\xi$ are considered. Bearing in mind that the transformations of $G_\xi$ are homogeneous linear orthogonal transformations, i.e., they are isomorphic to the three-dimensional subgroup of the Lorentz transformations (15), of the type

$$\xi_{i'} = D_{i'}^i \xi_i \qquad (D \equiv \det |D_{\xi'}^i| = \pm 1) \tag{22}$$

we have

$$d^3\xi = \text{sgn} \, \{D\} d^3\xi \tag{23}$$

and, as a matter of fact, $dP$ will be under the group $G_\xi$ of the internal transformations a pseudoscalar. Indeed, the invariant volume element of the local momentum space depends on the orientation of the basic $\lambda$-triad.

Due to the definition (17) of the group $G$ the relativistic phase-space volume-element in terms of the parametrization (18) can be given in the form

$$d\Omega \equiv m_0^3 \sqrt{-g} \, d^4 x \, d^3\xi \qquad (24)$$

being a pseudoscalar of the group $G$.[5]

The pseudoscalar character of the relativistic phase-space volume-element means that it can be oriented. Let it be denoted as a positive one if the underlying $\lambda$-triad is right-handed.

At a given instant of time, i.e., on the hyperplane $x^0 = $ const. of the coordinate space $\{x^\mu\}$, the phase-space volume element is reduced into the form

$$d\Omega_0 = m_0^3 \sqrt{-g} \, dx^1 \, dx^2 \, dx^3 \, d\xi_1 \, d\xi_2 \, d\xi_3 \equiv m_0^3 \sqrt{-g} \, d^3 x \, d^3\xi \qquad (25)$$

being the direct generalization of the well-known expression of the nonrelativistic gas theory. For sake of simple speaking $d\Omega_0$ may be called the *momentary expression* of the phase-space volume-element.

## 3. THE ZERO-POINT KINETIC ENERGY OF PERFECT FERION-GASES

In addition to the natural interest in the generalization of important physical concepts, the investigations to deal with the definitions of the energy and momentum of particles on the Fermi level in the case of relativistic fermion gases in Riemannian space–time continuum, i.e., in the case of an inhomogeneous but isotropic phase space, are also suggested by problems more recently raised in the neutrino astronomy.[6]

Having in mind the expressions (24) and (25) of the relativistic phase space volume element and its momentary value, respectively, all the nonrelativistic concepts and results may be generalized. First the somewhat more general case of fermions with nonvanishing rest mass ($m_0 \neq 0$) will be discussed, then the results will be specialized for the neutrino gas by the limiting process $m_0 \to 0$.

Owing to the classical results, the Fermi momentum and energy of the gaseous particles depend on the mean density $\bar{\rho}$ of fermions in the three-dimensional configuration space. This means that density can naturally be defined in a closed universe, but it can also be defined in the case of an open universe, keeping in mind the framework of infinite system of fermions, i.e., the limits $N \to \infty$ and $V \to \infty$ with the restriction $\bar{\rho} \equiv N/V = $ const.

Due to the spherical structure of the space–time continuum in the case of the steady-state solutions of the gravitational field equations and due to the dynamical isotropy of the system considered on the $x^0 = $ const. hyperplane, the spherical polar coordinates $\{r, \vartheta, \varphi\}$ and $\{p, \theta, \phi\}$ will, of course, be introduced in the coordinate space and in the local momentum case, respectively. This means that the momentary relativistic phase-space volume of the degenerate fermion gas considered has to be defined by

$$\Omega_0 \equiv \int_\alpha^R dr \int_0^\pi d\vartheta \int_0^{2\pi} d\varphi \int_0^{p_F} dp \int_0^\pi d\theta \int_0^{2\pi} d\varphi \sqrt{-g}\, p^2 \sin\theta \tag{26}$$

where the upper limit $R$ of the coordinate-space integral means either the radius of the universe or in the case of the spherical symmetric space–time domain taken into account before the limiting process $R \to \infty$ and the lower limit $\alpha$ is determined by its metrical properties. $p_F$ denotes the Fermi momentum.

The absolute value $p$ of the three-momentum component can be calculated from the normalization condition of the covariant four-momentum components

$$g^{\mu\nu}(x)p_\mu p_\nu = m_0^2 \tag{27}$$

since the momentum components with covariant transformation character have to be considered in the actual version of $dP$. This is also supported by the fact that the canonical four-momenta—as the derivatives of the Lagrangian according to the components of the four-velocity—are *per definitionem* covariant.

In the particular cases considered in the following, the components of the metrical fundamental tensor may generally be given as follows:

$$g_{00} = -h_{00}(r), \quad g_{11} = h_1(r), \quad g_{22} = h_2(r), \quad g_{33} = h_3(r)\sin^2\vartheta;$$
$$g_{\mu\nu} = 0 \,(\mu \neq \nu) \tag{28}$$

This means, however, that

$$\sqrt{-g} = \{h_0 h_1 h_2 h_3\}^{1/2} \sin\vartheta \tag{29}$$

and the normalization condition can be put in the form

$$-h_0^{-1}p_0^2 + \{h_1^{-1}p_1^2 + h_2^{-1}p_2^2 + h_3^{-1}\sin^{-2}\vartheta p_3^2\} = -m_0^2 \tag{30}$$

In fact, the definition of $p$ may finally be given by

$$p \equiv \{h_1^{-1}p_1^2 + h_2^{-1}p_2^2 + h_3^{-1}\sin^2\vartheta p_3^2\}^{1/2} = \{\epsilon^2 h_0^{-1}(x) - m_0^2\}^{1/2} \tag{31}$$

where it was taken into account that in our system of units $p_0 \equiv \epsilon$, denoting by $\epsilon$ the energy of the particles.

In order to introduce the inhomogeneous direction coordinates $\{\xi_i\}$ we have to normalizae the basic vectors of the $\lambda$-triad. In terms of the parametrization (18)—starting from the local rest frame of reference $\mathscr{K}^\circ$— it can be obtained on the one hand

$$\underset{1}{\lambda^1} = h_1^{-1/2}(r) \qquad \underset{2}{\lambda^2} = h_2^{-1/2}(r) \qquad \underset{3}{\lambda^3} = h_3^{-1/2}(r) \sin^{-1} \vartheta \qquad (32)$$

(all the other components of $\underset{i}{\lambda^\mu}$ are vanishing) and on the other

$$\xi_1 = \frac{h_1^{-1/2}(r)p_1}{m_0} \qquad \xi_2 = \frac{h_2^{-1/2}(r)p_2}{m_0} \qquad \xi_3 = \frac{h_3^{-1/2}(r)p_3}{m_0 \sin \vartheta} \qquad (33)$$

respectively. This means, however, that due to equation (31) the absolute value of the inhomogeneous direction coordinates

$$\xi = \frac{p}{m_0} = \left\{ \frac{\epsilon^2}{m_0^2} h_0^{-1}(r) - 1 \right\}^{1/2} \qquad (34)$$

and finally the momentary relativistic phase space volume element on the $x^0 = $ const. hyperplane can be written in the form

$$\Omega_0 = \frac{(4\pi)^2}{3} m_0^3 \int_\alpha^R dr \, \{h_0 h_1 h_2 h_3\}^{1/2} \left\{ \frac{\epsilon_F^2}{m_0^2} h_0^{-1}(r) - 1 \right\}^{3/2} \qquad (35)$$

where $\epsilon_F$ means the Fermi energy of the particles. Due to the definition of the three-dimensional volume in the coordinate space, we have

$$V \equiv 4\pi \int_\alpha^R dr \, \{h_0 h_1 h_2 h_3\}^{1/2} \qquad (36)$$

and bearing the characteristic property of the fermion gases (Pauli's principle) in mind, the number of the fermions $N$ can be written in the form

$$N = 2 \frac{(4\pi)^2}{3} \left( \frac{m_0}{2\pi\hbar} \right)^3 \int_\alpha^R dx \, \{h_0 h_1 h_2 h_3\}^{1/2} \left\{ \frac{\epsilon_F^2}{m_0^2} h_0^{-1}(r) - 1 \right\}^{3/2} \qquad (37)$$

where the familiar convention has been taken into account that the volume of the phase-space corresponding to each degree of freedom is just equal to the Planck's constant $h = 2\pi\hbar$.

The density $\rho(x)$ of the relativistic fermion gas on the hyperplain $x^0 = $ const. may be defined by

$$\bar{\rho} \equiv \frac{N}{V} = 4\pi \int\limits_{\alpha}^{R} dr^2 \rho(r) \tag{38}$$

Indeed, let the three-dimensional density $\rho(r)$ be introduced by

$$\rho(r) \equiv \frac{8\pi}{3V} \left(\frac{m_0}{2\pi\hbar}\right)^3 \frac{1}{r} \{h_0 h_1 h_2 h_3\}^{1/2} \left\{\frac{\epsilon_F^2}{m_0^2} h_0^{-1}(r) - 1\right\}^{3/2} \tag{39}$$

So far the Fermi energy is unknown and in terms of $\bar{\rho}$ is only implicitly determined by equations (37) and (38), respectively. In order to calculate it explicitly one has to carry out the $r$-integration in the equations above.

Finally, it seems to be worthwhile to introduce the three-dimensional energy density of the fermion gas being the corresponding $T_0^0(r)$ component of the energy-momentum tensor of the system. Due to the familiar definition of the zero-point kinetic energy this may be carried out by means of the relation.

$$E_0 = \int\limits_{\alpha}^{R} dr \int\limits_0^\pi d\vartheta \sin\vartheta \int\limits_0^{2\pi} d\varphi \int\limits_0^{\epsilon_F} d\epsilon\, r^2 \rho(r)\epsilon(r) = 2\pi\epsilon_F^2 \int\limits_{\alpha}^{R} dr\, r^2 \rho(r)$$

$$= 4\pi \int\limits_{\alpha}^{R} dr\, \{h_0 h_1 h_2 h_3\}^{1/2} T_0^0(r) \tag{40}$$

based on which the definition

$$T_0^0(r) \equiv \frac{4\pi}{3} \left(\frac{m_0}{2\pi\hbar}\right)^3 \frac{\epsilon_F^2}{V} \left\{\frac{\epsilon_F^2}{m_0^2} h_0^{-1}(r) - 1\right\}^{3/2} \tag{41}$$

can be suggested.

We are particularly interested in the special case of the relativistic neutrino gas. This means we have to carry out the limiting process $m_0 \to 0$ in the formulas above:

$$\bar{\rho} = \frac{4\epsilon_F^3}{3\pi V\hbar^3} \int\limits_{\alpha}^{R} dr\, h_0^{-1}\{h_1 h_2 h_3\}^{1/2} \tag{42}$$

i.e., the Fermi energy is given by

$$\epsilon_F = \hbar \left[\frac{3\pi}{4} \bar{\rho} \int\limits_{\alpha}^{R} dr\, \{h_0 h_1 h_2 h_3\}^{1/2}\right]^{1/3} \left[\int\limits_{\alpha}^{R} dr\, h_0^{-1}\{h_1 h_2 h_3\}^{1/2}\right]^{-1/3} \tag{43}$$

As to the three-dimensional density of the neutrino gas and to its three-dimensional energy density on the hyperplane $x^0 = $ const. we have

$$\rho(r) = \frac{\epsilon_F^3}{3\pi^2 V \hbar^3} \frac{1}{r^2 h_0} \{h_1 h_2 h_3\}^{1/2} \tag{44}$$

and

$$T_0^0(r) = \frac{\epsilon_F^5}{6\pi^2 V \hbar^3} h^{-3/2}(r) \tag{45}$$

respectively.

In order to obtain the final results in the case of different Riemannian space–time continuums, the explicit expressions of the metrical fundamental tensor have to be taken into account. These are summarized in Table I.

## Table I

Fermi Energy and Three-Dimensional Density and
Energy-Density Values in Different Riemannian
Space–Time Continuums

|  | Einstein's universe | Schwartzchild's solution | Special solution with cylindrical symmetry |
|---|---|---|---|
| $h_0$ | $1$ | $1 - \dfrac{\alpha}{r}$ | $1 - \dfrac{\alpha}{r}$ |
| $h_1$ | $\left(1 - \dfrac{r^2}{R^2}\right)^{-1}$ | $\left(1 - \dfrac{\alpha}{r}\right)^{-1}$ | $\left(1 - \dfrac{\alpha}{r}\right)^{-1}$ |
| $h_2 = h_3$ | $r^2$ | $r^2$ | $r^2\left(1 - \dfrac{\alpha}{r}\right)^{-1}$ |
| $\{h_1 h_2 h_3\}^{1/2}$ | $r^2\left(1 - \dfrac{r^2}{R^2}\right)^{-1/2}$ | $r^2\left(1 - \dfrac{\alpha}{r}\right)^{-1}$ | $r^4\left(1 - \dfrac{\alpha}{r}\right)^{-3/2}$ |
| $\alpha$ | $0$ | $2M$ | $2M$ |
| $\epsilon_F$ | $\hbar\left(\dfrac{3\pi}{4}\bar\rho\right)^{1/3}$ | $\hbar\left(\dfrac{3\pi}{4}\bar\rho\right)^{1/3}$ | $\hbar\left(\dfrac{3\pi}{4}\bar\rho\right)^{1/3}$ |
| $\epsilon_F(R)$ | — | $\hbar\left[\dfrac{3\pi}{4}\bar\rho\right]^{1/3}\left\{1 + \dfrac{9}{4}\dfrac{\alpha}{R}\right\}^{-1/3}$ | $\hbar\left(\dfrac{3\pi}{4}\bar\rho\right)^{1/3}\left\{1 + \dfrac{15}{4}\dfrac{\alpha}{R}\right\}^{-1/3}$ |
| $\rho(r)$ | $\dfrac{\epsilon_F^3}{3\pi^2 V \hbar^3}\left\{1 - \dfrac{r^2}{R^2}\right\}^{-1/2}$ | $\dfrac{\epsilon_F^3}{3\pi^2 V \hbar^3}\left[1 - \dfrac{\alpha}{r}\right]^{-3/2}$ | $\dfrac{\epsilon_F^3}{3\pi^2 V \hbar^3}\left\{1 - \dfrac{\alpha}{R}\right\}^{-5/2}$ |
| $T_0^0(r)$ | $\dfrac{\epsilon_F^5}{6\pi^2 V \hbar^3}$ | $\dfrac{\epsilon_F^5}{6\pi^2 V \hbar^2}\left\{1 - \dfrac{\alpha}{r}\right\}^{-3/2}$ | $\dfrac{\epsilon_F^5}{6\pi^2 V \hbar^3}\left\{1 - \dfrac{\alpha}{R}\right\}^{-3/2}$ |

## 4. REMARKS

Based on the proposed hypergeometrization of the relativistic phase-space formalism in the framework of the general line-elements geometry one can develop the relativistic kinetic theory of gases.

If we suppose that the function $f(x, \xi)$ in terms of the quantity

$$f(x, \xi)\, d\Omega \equiv -\sqrt{-g}\, f(x, \xi)\, d^4x\, d^3\xi \qquad (46)$$

being invariant of the group $G$ of the relativistic phase-space, gives the probability that the line element $\{x^\mu, \xi_i\}$ of the trajectory of the gaseous particle considered, i.e., its phase point with coordinates $\{x^\mu\}$ and with inhomogeneous direction coordinates $\{\xi_i\}$, can be found in the volume element $d\Omega$ of the relativistic phase-space, based on the stationarity condition

$$\delta \int_\Omega f(x, \xi)\, d\Omega = 0 \qquad (47)$$

and on the normalization equation

$$\int_\Omega f(x, \xi)\, d\Omega = 1 \qquad (48)$$

(where $\Omega$ denotes the domain of integration considered in the phase-space) the generalized Boltzmann's transport equation

$$\frac{\partial f}{\partial x^\mu} p^\mu + \frac{\partial f}{\partial \xi^i} \frac{d}{d\tau} \xi_i = - \left[ \frac{Df}{d\tau} \right]_{\text{coll}} \qquad (49)$$

can be obtained, where $\tau$ means the proper time and $[Df/d\tau]_{\text{coll}}$ the relativistic collision integral, respectively. It is easy to prove that (49) in special cases reduces into the well-known equation of Ken-Iti-Goto[7] and Chernikov,[8] respectively. Finally, all the interesting results of Israel's statistical thermodynamics[9] can be obtained.[5]

Finally, due to the fact that the suggested line-element geometry seems to be the natural geometrical background for Yukawa's bilocal theory of fields, we may hope that by improvement of our previous proposal[10] an adequate geometrical interpretation of the isospace — as internal degrees of freedom — and its transformations can be given. These problems and some details of the present paper will be published elsewhere.

## ACKNOWLEDGMENT

The author would like to express his sincere thanks to Professor Alladi Ramakrishnan for his kind invitation to participate in the Fifth Anniversary Symposium of the Institute of Mathematical Sciences and for his generous hospitality as well.

## REFERENCES

1. J. I. Horváth and A. Moór, *Koninkl. Ned. Akad. Wetenschap. Proc. Ser. A* **A58**: 421, 581 (1955).
2. J. I. Horváth, *Acta Phys.* **9**: 3 (1963).
3. J. I. Horváth, *Nuovo Cim. Suppl.* **9** (10): 444 (1958).
4. J. L. Synge, "Relativity: The General Theory," Amsterdam, North-Holland, 1960.
5. J. I. Horváth, *Wiss. Z. Friedrich-Schiller-Univ. Jena Math. Naturwiss Reihe.* **15**: 149 (1966).
6. L. Fodor, Zs. Kövesy, and G. Marx, *Acta Phys. Acad. Sci Hung.* **17**: 171 (1964).
7. Ken-Iti Goto, *Progr. Theoret. Phys.* (*Kyoto*) **20**: 1 (1958).
8. N. A. Chernikov, *Soviet Phys.* "*Doklady*" *English Transl* **2**: 248 (1957); **5**: 764, 786 (1960); **7**: 397, 414, 428 (1962); Reprint (Dubna, 1962).
9. W. Israel, *J. Math. Phys.* **4**: 1163 (1963).
10. J. I. Horváth, *Acta Phys. et Chem, Szeged,* **9**: 3 (1963).

# Variational Methods in Scattering Theory

CHARLES J. JOACHAIN

*UNIVERSITÉ LIBRE DE BRUXELLES‡*
**Brussels, Belgium**

---

## 1. INTRODUCTION

By writing the variational principle

$$\delta J = 0 \tag{1.1}$$

one expresses the fact that a certain functional $J$, characteristic of the system considered, is stationary for a "natural" system. This property leads to the well-known Euler–Lagrange equations of the system.

The aim of (direct) *variational methods* is to look for solutions of the variational problem (1.1) among trial functions which depend on a finite number of variational parameters $\alpha_1, \alpha_2, \ldots, \alpha_n$. The original functional $J$ becomes now a function $J(\alpha_1, \alpha_2, \ldots, \alpha_n)$ of the variational parameters and the stationary property (1.1) yields the system of equations

$$\frac{\partial J}{\partial \alpha_i} = 0 \qquad (i = 1, 2, \ldots, n) \tag{1.2}$$

to determine the "best" parameters $\alpha_1^0, \alpha_2^0, \ldots, \alpha_n^0$. Thus, by solving directly the variational problem in a restricted functional space, variational methods provide a powerful way of obtaining approximate solutions to the corresponding Euler–Lagrange equations, i.e., to a variety of differential and integral equations. By virtue of

---

‡Department of Theoretical Physics and Mathematics.

equation (1.2), the value of the functional at the "best" parameters, namely

$$J_0 = J(\alpha_1^0, \alpha_2^0, \ldots, \alpha_n^0)$$

is *correct to the first order* when variations are imposed on the trial functions considered. This last property is of great importance in performing practical calculations, the functional $J$ being thus relatively insensitive to errors made in the trial functions.

Since the early developments of quantum theory, variational methods proved to be very useful for solving bound state problems. It was only much later that the application of variational principles to quantum collision phenomena was proposed independently by Hulthen,[1] Tamm,[2] and Schwinger.[3] Numerous related methods and extensions have followed their work. The aim of all these methods, as applied to scattering processes, is to obtain variational principles for various quantities—such as phase shifts, elements of the $S$ matrix, the $T$ matrix, etc.—which characterize a collision.

It is convenient to classify variational principles into *differential* or *integral* forms. The various differential forms are based directly on differential equations and require trial functions satisfying the boundary conditions of the problem. In the case of integral forms, the Euler–Lagrange equations are integral equations and therefore the boundary conditions, being taken into account through the relevant Green's functions, need not be incorporated in the trial functions.

In Section 2, we shall study the Hulthén–Kohn[1,4,5] variational principle, which is currently the most widely used method based on the *differential* aspect of variationl principles for collision processes. Section 3 is devoted to the Schwinger variational principles,[6-9] relying on the corresponding integral equations. Finally, in Section 4, we present an introduction to the important problem of minimum principles in scattering theory.

## 2. THE HULTHÉN–KOHN VARIATIONAL PRINCIPLE

In this section, the Hulthén–Kohn method[1,4,5] for determining phase shifts and scattering amplitudes is analyzed. Related variational methods, also based on the differential aspect of variational principles, have been proposed[10] but will not be discussed here.

## 2.1. Variational Principles for Phase Shifts

Let us first consider the nonrelativistic scattering of a spinless particle of mass $m$ by a central potential $V(r)$. The initial and final propagator vectors are, respectively, $\mathbf{k}_i$ and $\mathbf{k}_f$ with $|\mathbf{k}_i| = |\mathbf{k}_f| = k$. For convenience, we introduce the reduced potential

$$U(r) = \frac{2m}{\hbar^2} V(r) \tag{2.1}$$

The exact wave function $\bar{\Psi}_i$ satisfies the Schrödinger equation

$$[\nabla^2 + k^2 - U]\bar{\Psi}_i = 0$$

and has the asymptotic behavior

$$\bar{\Psi}_i \xrightarrow[r \to \infty]{} e^{i\mathbf{k}_i \cdot \mathbf{r}} + \bar{f} \frac{e^{ikr}}{r} \tag{2.3}$$

We choose $\mathbf{k}_i \equiv 0z$ and expand $\bar{\Psi}_i$ in a series of Legendre polynomials

$$\bar{\Psi}_i(r, \theta) = r^{-1} \sum_{l=0}^{\infty} \bar{u}_l(r) P_l(\cos \theta) \tag{2.4}$$

where the function $\bar{u}_l(r)$ is the exact solution of the radial equation

$$L_l[\bar{u}_l] = \left[ \frac{d^2}{dr^2} + k^2 - \frac{l(l+1)}{r^2} - U(r) \right] \bar{u}_l(r) \equiv 0 \tag{2.5}$$

such that

$$\bar{u}_l(0) = 0 \tag{2.6}$$

At large distances, we "normalize" $\bar{u}_l(r)$ in such a way that

$$\bar{u}_l(r) \xrightarrow[r \to \infty]{} \sin(kr - \frac{l\pi}{2} + \bar{\eta}_l) \tag{2.7}$$

where $\bar{\eta}_l$ denotes the exact phase shift.

Now, let us consider the functional

$$I_l[u_l] = \int_0^{\infty} u_l(r) L_l[u_l(r)] dr \tag{2.8}$$

We note that $I_l = 0$ when $u_l = \bar{u}_l$. Let us adopt a trial function defined by

$$u_l(r) = \bar{u}_l(r) + \delta u_l(r) \tag{2.9}$$

and such that it satisfies the same boundary conditions as $\bar{u}_l(r)$, namely

$$u_l(0) = 0 \tag{2.10}$$

and

$$u_l(r) \xrightarrow[r \to \infty]{} \sin\left(kr - \frac{l\pi}{2} + \eta_l\right) \tag{2.11}$$

where $\eta_l$ is a trial phase shift. Hence

$$\delta u_l(0) = 0 \tag{2.12}$$

and

$$\delta u_l(r) \xrightarrow[r \to \infty]{} \cos\left(kr - \frac{l\pi}{2} + \eta_l\right)\delta\eta_l \tag{2.13}$$

We now calculate $\delta I_l$. Neglecting second-order terms and using Green's theorem, we get

$$\delta I_l = -k\,\delta\eta_l \tag{2.14}$$

so that the functional

$$J_l = I_l + k\eta_l \tag{2.15}$$

is stationary. Thus the variational phase shift $[\eta_l]$, correct to first order in $\delta u_l$, is given by

$$k[\eta_l] = k\eta_l + I_l \tag{2.16}$$

In actual calculations, one starts from a trial function $u_l(c_1, c_2, \ldots, c_n; \eta_l)$, satisfying the boundary conditions (2.10) and (2.11). One then calculates $J_l(c_1, c_2, \ldots, c_n; \eta_l)$ and the $(n+1)$ parameters are determined by the $(n+1)$ equations

$$\frac{\partial}{\partial c_i} J_l = 0 \quad \text{or} \quad \frac{\partial}{\partial c_i} I_l = 0 \quad (i = 1, 2, \ldots, n) \tag{2.17}$$

and

$$\frac{\partial}{\partial \eta_l} J_l = 0 \quad \text{or} \quad \frac{\partial}{\partial \eta_l} I_l = -k \tag{2.18}$$

The trial phase shift $\eta_l$ being determined in this way, the "variationally correct" phase shift $[\eta_l]$ is obtainined from equation (2.16).

Slightly different variational principles can be obtained with different "normalizations" for the functions $\bar{u}_l(r)$ and $u_l(r)$. For example, replacing equations (2.7) and (2.11), respectively, by

$$\bar{u}_l(r) \xrightarrow[r \to \infty]{} \sin\left(kr - \frac{l\pi}{2}\right) + \bar{\lambda}_l \cos\left(kr - \frac{l\pi}{2}\right) \tag{2.19}$$

and

$$u_l(r) \xrightarrow[r \to \infty]{} \sin\left(kr - \frac{l\pi}{2}\right) + \lambda_l \cos\left(kr - \frac{l\pi}{2}\right) \tag{2.20}$$

where $\bar{\lambda} = \tan \bar{\eta}_l$ and $\lambda_l = \tan \eta_l$, one gets

$$\delta(I_l + k\lambda_l) = 0 \tag{2.21}$$

In this case the equation

$$k[\lambda_l] = k\lambda_l + I_l \tag{2.22}$$

replaces equation (2.16). The variational principle (2.21) is more advantageous to use in practice than the corresponding variational principle (2.14) because the quantity $\lambda_l$, directly related to the phase shift, appears as a *linear* parameter in the trial function. The Hulthén–Kohn procedure leading to equations (2.17) and (2.18) is unchanged, with now $J_l = I_l + k\lambda_l$ and thus

$$\frac{\partial}{\partial c_i} I_l = 0 \tag{2.23}$$

and

$$\frac{\partial}{\partial \lambda_l} I_l = -k \tag{2.24}$$

This procedure is preferable to the original method of Hulthén[1] in which the variational parameters $c_1, c_2, \ldots, c_n$ and $\lambda_l$ are determined by the equations

$$\frac{\partial}{\partial c_i} I_l = 0 \qquad (i = 1, 2, \ldots, n) \tag{2.25}$$

and

$$I_l(c_1, c_2, \ldots, c_n; \lambda_l) = 0 \tag{2.26}$$

The reason is that the last equation (2.26) is quadratic in $\lambda_l$, thus introducing an ambiguity in the choice of $\lambda_l$. Another reason to prefer the Hulthén–Kohn procedure in certain circumstances will be given in Section 4.

As an example, we illustrate the use of the Hulthén–Kohn variational prinicple (2.21)–(2.22) for $S$-state electron scattering by the static fields of hydrogen[11] and helium[12] for which

$$U(r) = -2N\left(Z' + \frac{1}{r}\right)e^{-2Z'r} \tag{2.27}$$

where $N = Z' = 1$ for hydrogen and $N = 2, Z' = 27/16$ for helium. Lengths are in units of $a_0 = \hbar^2/me^2$, where $m$ is the electron mass.

The trial function, satisfying the boundary conditions (2.10) and (2.20), is chosen to be

$$u_0(r) = \sin kr + (\lambda_0 + ce^{-Z'r})(1 - e^{-Z'r}) \cos kr \qquad (2.28)$$

Results are displayed in Tables I and II for several values of $ka_0$, and compared with a direct numerical integration of the original Schrödinger equation. The agreement is surprisingly good, given the simplicity of the trial function (2.28).

### Table I

Comparison of Phase Shifts $\eta_0$ for $S$-Wave Scattering of an Electron by the Static Field of an Hydrogen atom, Calculated by Numerical Integration and by the Hulthén-Kohn Variational Method. Taken from Ref. 11.

| $ka_0$ | Phase shift $\eta_0$ (in radians) | |
|---|---|---|
| | Numerical integration | Hulthén-Kohn variational method |
| 0.1 | 0.730 | 0.721 |
| 0.3 | 1.046 | 1.045 |
| 0.5 | 1.045 | 1.044 |
| 1.0 | 0.906 | 0.904 |

### Table II

Comparison of Phase Shifts $\eta_0$ for $S$-Wave Scattering of an Electron by the Static Field of an Helium Atom, Calculated by Numerical Integration and by the Hulthén-Kohn Variational Method. Taken from Ref. 12.

| $ka_0$ | Phase shift $\eta_0$ (in radians) | |
|---|---|---|
| | Numerical integration | Hulthén-Kohn variational method |
| 0.136 | 2.57 | 2.34 |
| 0.608 | 1.66 | 1.58 |
| 1.053 | 1.36 | 1.34 |
| 1.922 | 1.09 | 1.07 |

The Hulthén-Kohn method can easily be extended to the calculation of phase shifts corresponding to the elastic scattering of a particle by a composite system[11,13] including, if necessary, the effects of the Pauli exclusion principle. As a classic example, let us consider the elastic scattering of electrons from the ground state of hydrogen atoms. The Hamiltonian of the system is

$$H = -\frac{\hbar^2}{2m}(\nabla_1^2 + \nabla_2^2) - \frac{e^2}{r_1} - \frac{e^2}{r_2} + \frac{e^2}{r_{12}} \qquad (2.29)$$

where $m$ is the electron mass. The trial wave functions $\Psi_i^\pm$—the superscripts refer, respectively, to the singlet (space symmetric) and the triplet (space antisymmetric) cases—may be given the following asymptotic form:

$$\Psi_i^\pm(\mathbf{r}_1, \mathbf{r}_2) \xrightarrow[\substack{r_1 \to \infty \\ r_2 \to \infty}]{} \frac{1}{\sqrt{4\pi}}\sqrt{2}\; e^{-r_2}\frac{1}{kr_1}\left[\sin\left(kr_1 - \frac{l\pi}{2}\right) + \tan\eta_i^\pm\right.$$

$$\left. \cos\left(kr_1 - \frac{l\pi}{2}\right)\right] \cdot Y_{l,0}(\theta_2) \pm (1 \leftrightarrow 2) \qquad (2.30)$$

where lengths are in units of the Bohr radius $a_0$. A straightforward extension of the analysis leading to equations (2.21) and (2.22) yields the variational principle

$$\left[\frac{\tan\eta_i^\pm}{k}\right] = \frac{\tan\eta_i^\pm}{k} + \frac{2m}{\hbar^2}\int \Psi_i^\pm(E - H)\Psi_i^\pm\, d\mathbf{r}_1\, d\mathbf{r}_2 \qquad (2.31)$$

Detailed applications of this last formula will be discussed in Section 2.3.

## 2.2. Variational Principle for the Scattering Amplitude

The Hulthén–Kohn variational principle can also be formulated for the scattering amplitude.[5] In this case, we start from the Schrödinger equation (2.2) and consider the integral, similar to that of equations (2.8)

$$I[\Psi_{-\mathbf{k}_f}, \Psi_{\mathbf{k}_i}] = \int \Psi_{-\mathbf{k}_f}(\mathbf{r})[\nabla^2 + k^2 - U(\mathbf{r})]\Psi_{\mathbf{k}_i}(\mathbf{r}) \qquad (2.32)$$

where the wave function $\Psi_{\mathbf{k}_i}$ has the asymptotic form (2.3) and $\Psi_{-\mathbf{k}_f}$ has a similar asymptotic behavior with $\mathbf{k}_i$ replaced by $(-\mathbf{k}_f)$. Introducing the variations $\delta\Psi_{\mathbf{k}_i}$ and $\delta\Psi_{\mathbf{k}_f}$ such that

$$\delta\Psi_{\mathbf{k}_i} \xrightarrow[r \to \infty]{} \frac{e^{ikr}}{r}\,\delta f(\mathbf{k}_i \cdot \hat{r}) \qquad (2.33a)$$

and

$$\delta\Psi_{\mathbf{k}_f} \xrightarrow[r \to \infty]{} \frac{e^{ikr}}{r}\,\delta f(-\mathbf{k}_f \cdot \hat{r}) \qquad (2.33b)$$

one obtains, after some manipulations[5]

$$\delta I = -4\pi \, \delta f(\mathbf{k}_i \cdot \mathbf{k}_f) \tag{2.34}$$

so that the functional

$$J \equiv I + 4\pi f(\mathbf{k}_i \cdot \mathbf{k}_f) \tag{2.35}$$

is stationary against variations of $\Psi_{\mathbf{k}_i}$ and $\Psi_{-\mathbf{k}_f}$ around their correct values. Since $I = 0$ for the exact wave functions, we obtain

$$[f(\mathbf{k}_i \cdot \mathbf{k}_f)] = f(\mathbf{k}_i \cdot \mathbf{k}_f) + \frac{1}{4\pi} \, I[\Psi_{-\mathbf{k}_f}, \Psi_{\mathbf{k}_i}] \tag{2.36}$$

which is a direct extension of equation (2.15). The variational parameters $c_i(i = 1, 2, \ldots, n)$ in the trial functions are now determined by the conditions

$$\frac{\partial}{\partial c_i} \, (I + 4\pi f) = 0 \qquad (i = 1, 2, \ldots, n) \tag{2.37}$$

Note that if we choose the simplest possible trial functions, namely,

$$\Psi_{\mathbf{k}_i} = e^{i\mathbf{k}_i \cdot \mathbf{r}}$$

and

$$\Psi_{\mathbf{k}_f} = e^{i\mathbf{k}_f \cdot \mathbf{r}} \tag{2.38}$$

we have $f = 0$ and

$$I = -\int e^{i(\mathbf{k}_i - \mathbf{k}_f) \cdot r} U(r) d\mathbf{r} \tag{2.39}$$

so that $[f]$, as given by the variational principle (2.36), reduces to the first Born approximation. Further generalization of this method have been proposed by Kohn[5] to determine the eigenphase shifts of the $S$-matrix.

## 2.3. The Convergence Problems

We now turn to a more detailed analysis of the Hulthén–Kohn method for determining phase shifts. We are especially concerned with the convergence of the method when an increased number of variational parameters is used in the trial function. We shall follow closely the work of Schwartz[14] and illustrate the general procedure in the realistic case of $S$-wave electron-hydrogen elastic scattering. The Hulthén–Kohn variational principle (2.31) for the quantity $\lambda = \tan \eta_0^{\pm}/k$ then gives

$$[\lambda] = \lambda + \frac{2m}{\hbar^2} \int \psi (E - H) \psi d\tau \tag{2.40}$$

The trial function $\psi$ is constructed as follows:

$$\psi = \varphi + \sum_{i=1}^{N} c_i \chi_i \qquad (2.41)$$

where the asymptotic part is

$$\varphi = \frac{1}{4\pi}\sqrt{2}\, e^{-r_2}\left[\frac{\sin kr_1}{kr_1} + \lambda\,\frac{\cos kr_1}{r_1}\left(1 - e^{-\kappa r_1/2}\right)\right] \pm (1 \leftrightarrow 2) \qquad (2.42)$$

and the basis functions $\chi_i$ are given by

$$\chi_{lmn} = e^{-\kappa(r_1+r_2)/2}\frac{(r_1^m r_2^n \pm r_1^n r_2^m)r_{12}^l}{4\pi\sqrt{2}} \qquad (2.43)$$

The additional nonlinear parameter $\kappa$ is a scale factor to be varied at the end of the calculations. The basis functions $\chi_{lmn}$ are ordered in such a way that one always uses all functions with $l + m + n \leq N$, and then increases $N$.

For a given value of the scale factor $\kappa$, the problem now reduces to a simple one of matrix inversion. Indeed, defining

$$M_{ij} = M_{ji} = \frac{2m}{\hbar^2}\int \chi_i(E - H)\chi_j\,d\tau \qquad (2.44)$$

$$R_i = \frac{2m}{\hbar^2}\int \chi_i(E - H)\varphi\,d\tau \qquad (2.45)$$

and

$$B = \frac{2m}{\hbar^2}\int \varphi(E - H)\varphi\,d\tau \qquad (2.46)$$

the variational principle (2.40) yields

$$[\lambda] = \sum_{i,j=1}^{N} c_i c_j M_{ij} + 2\sum_{i=1}^{N} c_i R_i + B + \lambda \qquad (2.47)$$

Variation of the constants then gives the system of $N$ linear equations.

$$\sum_{j=1}^{N} M_{ij}c_j = -R_i \qquad (i = 1, 2, \dots, n) \qquad (2.48)$$

so that, by inverting the matrix $(M)$, one obtains the variational parameters $c_i$. Having thus determined the parameters $c_i$, it is an easy matter to vary by respect to $\lambda$, thus determining the stationary value $[\lambda]$, "second-order accurate," after substitution in equation (2.47).

Let us now analyze equation (2.48). The operator $(E - H)$ which enters in the definition of the matrix $(M)$ has a continuum of eigenvalues passing through the value zero. The finite matrix $(M)$, by which we represent $(E - H)$ in the space spanned by our $N$ trial functions $\chi_i$, has $N$ eigenvalues and therefore, occasionally, can have an eigenvalue very close to zero. In this case the whole variational procedure is meaningless.

For example if we return to the electron-hydrogen scattering problem and let the scale factor $\kappa$ vary in (2.42) and (2.43), one can expect one or more of the eigenvalues of $(M)$ to be close to zero for certain ranges of $\kappa$. The saving grace of the variational procedure is that as $N$ increases, the singularities appear to be less strong and that the average value of $[\lambda]$ between the singularities becomes smooth and flat. In this way a reasonably convergent extrapolation procedure can be devised and precise values of the phase shifts can be obtained. Table III summarizes the results of Schwartz[15] for electron-hydrogen scattering. These represent probably the most accurate results up to date for a real three-body problem.

## Table III

Results of Variational Calculations of $S$-Wave Electron-Hydrogen Scattering, Using the Hulthén-Kohn Method; $S = 0$ Refers to the Singlet $S = 1$ to the Triplet Case. The Numbers in Parentheses Give the Uncertainty in the Last Digit Quoted. Taken from Ref. 15.

| $ka_0$ | Phase shift $\eta_0$ (in radians) | |
|---|---|---|
| | $S = 0$ | $S = 1$ |
| 0.1 | 2.553(1) | 2.9388(4) |
| 0.3 | 1.6964(5) | 2.4996(8) |
| 0.5 | 1.202(1) | 2.1046(4) |
| 0.7 | 0.930(1) | 1.7797(6) |

## 3. THE SCHWINGER VARIATIONAL PRINCIPLE

In this section we want to study variational principles based directly on the integral equations of scattering phenomena. We shall proceed by using the methods of formal scattering theory.

### 3.1. Preliminaries

Let us consider a general collision and assume that the total Hamiltonian of the system can be decomposed in the initial channel as

$$H = H_i + V_i \tag{3.1}$$

where $V_i$ is the interaction between the colliding systems and where

$$H_i \Phi_i = E_i \Phi_i \tag{3.2}$$

$\Phi_i$ being the "free" wave describing the system when the colliding particles are far apart. In the same way, in the final channel, we write

$$H = H_f + V_f \tag{3.2b}$$

and

$$H_f \Phi_f = E_f \Phi_f \tag{3.1b}$$

Let $\langle f|S|i \rangle$ be the $S$-matrix element describing the collision considered. We have[16]

$$\langle f|S|i \rangle = \delta_{if} - 2\pi i \, \delta(E_i - E_f)\delta(\mathbf{P}_i - \mathbf{P}_f)T_{if} \tag{3.3}$$

The vectors $\mathbf{P}_i$ and $\mathbf{P}_f$ are the initial and final eigenvalues of the total momentum operator $\mathscr{P}$, i.e.,

$$\mathscr{P}\Phi_i = \mathbf{P}_i \Phi_i \tag{3.4a}$$

and

$$\mathscr{P}\Phi_f = \mathbf{P}_f \Phi_f \tag{3.4b}$$

and $T_{if}$ is the transition matrix element on the energy-momentum shell such that

$$T_{if} = \langle \Phi_f|V_f|\Psi_i^{(+)} \rangle = \langle \Psi_f^{(-)}|V_i|\Phi_i \rangle \tag{3.5}$$

In the equation (3.5), the objects $\Psi_i^{(+)}$ and $\Psi_f^{(-)}$ denote the solutions of the Lippmann–Schwinger equations

$$\Psi_i^{(\pm)} = \Phi_i + G_i^{(\pm)} V_i \Psi_i^{(\pm)} \tag{3.6a}$$

and

$$\Psi_f^{(\pm)} = \Phi_f + G_f^{(\pm)} V_f \Psi_f^{(\pm)} \tag{3.6b}$$

where

$$G_i^{(\pm)} = \lim_{\epsilon \to 0^+} \frac{1}{E_i - H_i \pm i\epsilon} \tag{3.7a}$$

and

$$G_f^{(\pm)} = \lim_{\epsilon \to 0^+} \frac{1}{E_f - H_f \pm i\epsilon} \tag{3.7b}$$

are the Green operators associated, respectively, with $H_i$ and $H_f$

and acting in the barycentric subspace of the system. Equations (3.6) have the well-known formal solutions

$$\Psi_i^{(\pm)} = \Phi_i + G^{(\pm)} V_i \Phi_i \tag{3.8a}$$

and

$$\Psi_f^{(\pm)} = \Phi_f + G^{(\pm)} V_f \Phi_f \tag{3.8b}$$

where the Green operators $G^{(\pm)}$ associated to $H$ satisfy the equations

$$G^{(\pm)} = G_i^{(\pm)} + G_i^{(\pm)} V_i G^{(\pm)} \tag{3.9a}$$

and

$$G^{(\pm)} = G_f^{(\pm)} + G_f^{(\pm)} V_f G^{(\pm)} \tag{3.9b}$$

The Moller wave operators $\Omega_i^{(\pm)}$ and $\Omega_f^{(\pm)}$ are given by

$$\Omega_i^{(\pm)} = 1 + G^{(\pm)} V_i \tag{3.10a}$$

and

$$\Omega_f^{(\pm)} = 1 + G^{(\pm)} V_f \tag{3.10b}$$

Hence

$$\Psi_i^{(\pm)} = \Omega_i^{(\pm)} \Phi_i \tag{3.11a}$$

and

$$\Psi_f^{(\pm)} = \Omega_f^{(\pm)} \Phi_f \tag{3.11b}$$

We shall also need the transition operator

$$T^{if} = V_f + V_f G^{(+)} V_i \tag{3.12}$$

such that

$$\langle f|T|i \rangle = \langle \Phi_f | T^{if} | \Phi_i \rangle \tag{3.13}$$

From (3.10a) and (3.10b), we have (denoting Hermitian conjugation by a dagger)

$$V_f [\Omega_i^{(+)} - 1] = [\Omega_f^{(-)} - 1]^\dagger V_i \tag{3.14}$$

while, from equations (3.9) and (3.10), we get

$$[1 - G_i^{(\pm)} V_i] \Omega_i^{(\pm)} = 1 \tag{3.15}$$

and

$$[1 - G_f^{(\pm)} V_f] \Omega_f^{(\pm)} = 1 \tag{3.16}$$

A little algebra then gives the identities[8]

$$[1 - G_f^{(+)} V_f][\Omega_i^{(+)} - 1] = G_f^{(+)} V_i \tag{3.17}$$

and

$$[\Omega_f^{(-)} - 1]^{\dagger}[1 - V_i G_i^{(+)}] = V_f G_i^{(+)} \tag{3.18}$$

Multiplying equation (3.17) to the left by $\Omega_f^{(-)\dagger} V_f$, we obtain

$$\Omega_f^{(-)\dagger}[V_f - V_f G_f^{(+)} V_f][\Omega_i^{(+)} - 1] = \Omega_f^{(-)\dagger} V_f G_f^{(+)} V_i \tag{3.19}$$

while, upon multiplying equation (3.18) to the right by $V_i \Omega_i^{(+)}$, we have

$$[\Omega_f^{(-)} - 1]^{\dagger}[V_i - V_i G_i^{(+)} V_i]\Omega_i^{(+)} = V_f G_i^{(+)} V_i \Omega_i^{(+)} \tag{3.20}$$

Finally, we note that the transition operator defined by equation (3.12) can be written with the help of (3.10a)

$$T^{if} = V_f \Omega_i^{(+)} \tag{3.21}$$

or else, by making use of (3.14)

$$T^{if} = \Omega_f^{(-)\dagger} V_i + (V_f - V_i) \tag{3.22}$$

## 3.2. Variational Principle for the $T$-Matrix

With these preliminary formulas established, it is now a simple matter to obtain variational principles for the transition matrix element $T_{if}$. Let us consider first the expression

$$[R_1] = \Omega_f^{(-)\dagger} V_f G_f^{(+)} V_i + V_f \Omega_i^{(+)} - \Omega_f^{(-)\dagger}(V_f - V_f G_f^{(+)} V_f)$$
$$(\Omega_i^{(+)} - 1) \tag{3.23}$$

Using (3.19) and (3.20), we see that its exact value is given by

$$[R_1] = T^{if} \tag{3.24}$$

Moreover, this expression is stationary for independent variations of the Moller wave operators around their correct values. Indeed,

$$\delta[R_1] = \delta\Omega_f^{(-)\dagger} V_f\{G_f^{(+)} V_i - (1 - G_f^{(+)} V_f)(\Omega_i^{(+)} - 1)\}$$
$$+ \{1 - \Omega_f^{(-)\dagger}(1 - V_f G_f^{(+)})\}V_f \delta\Omega_i^{(+)} \tag{3.25}$$

and from (3.17) and (3.16)

$$\delta[R_1] = 0 \tag{3.26}$$

Thus the expression (3.23) provides a variational principle for the operator $T^{if}$. A similar variational principle, involving only the Green operator $G_i^{(+)}$, is given by

$$[R_2] = V_f G_i^{(+)} V_i \Omega_i^{(+)} + \Omega_f^{(-)\dagger} V_i + (V_f - V_i)$$
$$- (\Omega_f^{(-)} - 1)^{\dagger}(V_i - V_i G_i^{(+)} V_i)\Omega_i^{(+)} \tag{3.27}$$

Indeed, using (3.22) and (3.20), the exact value of this expression is

$$[R_2] = T^{if} \tag{3.28}$$

and, from (3.18) and (3.16)

$$\delta[R_2] = 0 \tag{3.29}$$

Taking matrix elements of the stationary expressions (3.23) and (3.27) between two "free" states $\Phi_i$ and $\Phi_f$ of equal energy and making use of equations (3.11), we obtain for the transition matrix elements the variational principles

$$[T_{if}] = \langle \Psi_f^{(-)} | V_f \{ 1 + G_f^{(+)} (V_i - V_f) \} | \Phi_i \rangle + \langle \Phi_f | V_f | \Psi_i^{(+)} \rangle$$
$$- \langle \Psi_f^{(-)} | V_f - V_f G_f^{(+)} V_f | \Psi_i^{(+)} \rangle \tag{3.30}$$

and

$$[T_{if}] = \langle \Phi_f | \{ 1 + (V_f - V_i) G_i^{(+)} \} V_i | \Psi_i^{(+)} \rangle + \langle \Psi_f^{(-)} | V_i | \Phi_i \rangle$$
$$- \langle \Psi_f^{(-)} | V_i - V_i G_i^{(+)} V_i | \Psi_i^{(+)} \rangle \tag{3.31}$$

which are bilinear expressions in the wave functions $\Psi_i^{(+)}$ and $\Psi_f^{(-)}$. The fractional form of the variational principle is obtained by substituting

$$\Psi_i^{(+)} \longrightarrow A \Psi_i^{(+)} \qquad \Psi_f^{(-)} \longrightarrow B \Psi_f^{(-)} \tag{3.32}$$

in (3.30) and (3.31) and considering the amplitudes $A$ and $B$ as variational parameters. We obtain in this way

$$[T_{if}] = \frac{\langle \Phi_f | V_f | \Psi_i^{(+)} \rangle \langle \Psi_f^{(-)} | V_f \{ 1 + G_f^{(+)} (V_i - V_f) \} | \Phi_i \rangle}{\langle \Psi_f^{(-)} | V_f - V_f G_f^{(+)} V_f | \Psi_i^{(+)} \rangle} \tag{3.33}$$

and

$$[T_{if}] = \frac{\langle \Psi_f^{(-)} | V_i | \Phi_i \rangle \langle \Phi_f | \{ 1 + (V_f - V_i) G_i^{(+)} \} V_i | \Psi_i^{(+)} \rangle}{\langle \Psi_f^{(-)} | V_i - V_i G_i^{(+)} V_i | \Psi_i^{(+)} \rangle} \tag{3.34}$$

In the case of direct collisions (elastic or inelastic, but no rearrangement) for which $V_i = V_f = V$, $H_i = H_f = H$, and $G_i^{(\pm)} = G_f^{(\pm)} = G_0^{(\pm)}$ the bilinear forms (3.30) and (3.31) of the variational principles reduce to

$$[T_{if}] = \langle \Psi_f^{(-)} | V | \Phi_i \rangle + \langle \Phi_f | V | \Psi_i^{(+)} \rangle$$
$$- \langle \Psi_f^{(-)} | V - V G_0^{(+)} V | \Psi_i^{(+)} \rangle \tag{3.35}$$

and the fractional forms (2.33) and (2.34) become

$$[T_{if}] = \frac{\langle \Psi_f^{(-)} | V | \Phi_i \rangle \langle \Phi_f | V | \Psi_i^{(+)} \rangle}{\langle \Psi_f^{(-)} | V - V G_0^{(+)} V | \Psi_i^{(+)} \rangle} \tag{3.36}$$

    Similar variational principles can be formulated for the reaction matrix.[7] Thus, for example in the case of direct collisions, we have

$$[R_{if}] = \frac{\langle \Psi_f | V | \Phi_i \rangle \langle \Phi_f | V | \Psi_i \rangle}{\langle \Psi_f | V - V G_0^0 V | \Psi_i \rangle} \tag{3.37}$$

where

$$\Psi_i = \Phi_i + G_0^0 V \Psi_i \tag{3.38a}$$
$$\Psi_f = \Phi_f + G_0^0 V \Psi_f \tag{3.38b}$$

and

$$G_0^0 = P \left( \frac{1}{E - H_0} \right) \tag{3.39}$$

where $P$ denotes the Cauchy principal value.

    As a first illustration, let us replace in the fractional expressions (3.33) and (3.34) the unknown wave functions $\Psi_i^{(+)}$ and $\Psi_f^{(-)}$ by the Born free waves, namely,

$$\Psi_i^{(+)} \simeq \Phi_i \tag{3.40a}$$

and

$$\Psi_f^{(-)} \simeq \Phi_f \tag{3.40b}$$

We obtain the following approximate expressions (which we denote by the subscript $B$):

$$[T_{if}]_B = T_{if}^{(B)} \left[ 1 + \frac{\langle \Phi_f | V_f G_f^{(+)} V_i | \Phi_i \rangle}{\langle \Phi_f | V_f - V_f G_f^{(+)} V_f | \Phi_i \rangle} \right] \tag{3.41}$$

and

$$[T_{if}]_B = T_{if}^{(B)} \left[ 1 + \frac{\langle \Phi_f | V_f G_i^{(+)} V_i | \Phi_i \rangle}{\langle \Phi_f | V_i - V_i G_i^{(+)} V_i | \Phi_i \rangle} \right] \tag{3.42}$$

where

$$T_{if}^{(B)} = \langle \Phi_f | V_f | \Phi_i \rangle = \langle \Phi_f | V_i | \Phi_i \rangle \tag{3.43}$$

is the transition matrix element in first Born approximation.

    It is interesting to compare the formulas (3.40) and (3.41) with the Born development. If, in the right-hand side, the second term of the bracket—whose numerator is of the second order in the interaction potentials—is neglected, one recovers the first Born approximation. However, if this term is maintained but the second-order part of the dominator is neglected, one finds the second Born approximation.

The situation is even simpler in the case of direct collisions. We then have, with the approximation (3.40)

$$[T_{if}]_B = \frac{|\langle \Phi_f | V | \Phi_i \rangle|^2}{\langle \Phi_f | V - V G_0^{(+)} V | \Phi_i \rangle} \qquad (3.44)$$

If we expand formula (3.44) in powers of $V$, we get

$$[T_{if}]_B = \langle \Phi_f | V | \Phi_i \rangle + \langle \Phi_f | V G_0^{(+)} V | \Phi_i \rangle + \cdots \qquad (3.45)$$

which agrees with the Born series through second order in the interaction potential. Note that equations (3.41), (3.42), and (3.44) are expected to give more accurate results than the second Born approximation, since they are based on a variational principle. It is worth noting also that the approximation (3.40) made in the Hulthén-Kohn variational principle (2.36) only gave the first Born approximation.

An evaluation of (3.36) using the approximation (3.40) together with more elaborate choices has been carried out for $^3S_1$ nucleon-nucleon scattering described by a Yukawa potential[17] or by an exponential or a gaussian potential.[18] Since the Green's function $G_0^{(+)}(\mathbf{r}, \mathbf{r}')$ is given in this case by

$$G_0^{(+)}(\mathbf{r}, \mathbf{r}') = -\frac{2\mu}{\hbar^2} \frac{e^{ik|\mathbf{r} - \mathbf{r}'|}}{4\pi|\mathbf{r} - \mathbf{r}'|} \qquad (3.46)$$

where $\mu$ is the reduced mass of the colliding systems, the variational principle (3.44) can be written explicitly as

$$[T_{if}]_B = \frac{\left[\int e^{i(\mathbf{k}_i - \mathbf{k}_f) \cdot \mathbf{r}} V(\mathbf{r}) d\mathbf{r}\right]^2}{\int e^{i(\mathbf{k}_i - \mathbf{k}_f) \cdot \mathbf{r}} V(\mathbf{r}) d\mathbf{r} + \int d\mathbf{r} d\mathbf{r}' e^{-i\mathbf{k}_f \cdot \mathbf{r}} V(\mathbf{r}) \left(\frac{\mu e^{ik|r - r'|}}{2\pi\hbar^2|\mathbf{r} - \mathbf{r}'|}\right) V(r') e^{i\mathbf{k}_i \mathbf{r}'}} \qquad (3.47)$$

The variational results turn out indeed to be always superior to the second Born approximation. This statement, however, is meaningless at low energies, for which the Born series does not converge in this case. Actually, with the simple choice (3.40) used in the variational expression (3.47), the first Born approximation may even give less inaccurate cross sections than the variational estimate. Such examples illustrate the difficulties that may appear when poor trial functions are used in variational principles. Application of the variational expression (3.44) to weaker interactions provide in fact sufficiently accurate scattering amplitudes.[19]

As another application of the variational principle (3.36) for direct collisions, we consider the non-relativistic scattering of a particle of mass $\mu$ by a separable nonlocal potential[18,20] of the form

$$\langle \mathbf{r}|V|\mathbf{r}'\rangle = \lambda_0 v(\mathbf{r})v(\mathbf{r}') \tag{3.46}$$

where $\lambda_0$ is a real parameter. The exact scattering amplitude can be obtained easily[21] and may be written as

$$\bar{f} = -\frac{\mu}{2\pi\hbar^2}\lambda_0 \frac{\int e^{-i\mathbf{k}_f\cdot\mathbf{r}}v(\mathbf{r})d\mathbf{r}\int v(\mathbf{r}')e^{i\mathbf{k}_i\cdot\mathbf{r}'}\,d\mathbf{r}'}{1+\frac{\mu}{2\pi\hbar^2}\lambda_0\int d\mathbf{r}\,d\mathbf{r}'v(\mathbf{r})\frac{e^{ik|\mathbf{r}-\mathbf{r}'|}}{|\mathbf{r}-\mathbf{r}'|}v(\mathbf{r}')} \tag{3.47}$$

On the other hand, from the fractional form (3.44) of Schwinger's variational principle, we have

$$[f] = -\frac{\mu}{2\pi\hbar^2}\frac{A\cdot B}{C} \tag{3.48}$$

where

$$A = \lambda_0\int \Psi_f^{(-)*}(\mathbf{r})v(\mathbf{r})d\mathbf{r}\int v(\mathbf{r}')e^{i\mathbf{k}_i\cdot\mathbf{r}}d\mathbf{r}' \tag{3.49}$$

$$B = \lambda_0\int e^{-i\mathbf{k}_f\cdot\mathbf{r}_1}v(\mathbf{r}_1)d\mathbf{r}_1\int v(\mathbf{r}_1')\Psi_i^{(+)}(\mathbf{r}_1')d\mathbf{r}_1' \tag{3.50}$$

and

$$C = \lambda_0\int \Psi_f^{(-)*}(\mathbf{r})v(\mathbf{r})d\mathbf{r}\int v(\mathbf{r}')\Psi_i^{(+)}(\mathbf{r}')d\mathbf{r}'$$
$$+ \frac{\mu}{2\pi\hbar^2}\lambda_0^2\int \Psi_f^{(-)*}(\mathbf{r})v(\mathbf{r})\,d\mathbf{r}\iint v(\mathbf{r}')\frac{e^{ik|\mathbf{r}'-\mathbf{r}_1|}}{|\mathbf{r}'-\mathbf{r}_1|}v(\mathbf{r}_1)d\mathbf{r}'\,d\mathbf{r}_1\int v(\mathbf{r}_1')\Psi_i^{(+)}(\mathbf{r}_1')d\mathbf{r}_1' \tag{3.51}$$

Thus, after simplification

$$[f] = \bar{f} \tag{3.52}$$

for any choice of trial functions $\Psi_i^{(+)}$ and $\Psi_f^{(-)}$ and for any separable potential.[18-20]

An interesting modification of the Schwinger variational principle for the $T$-matrix has recently been proposed by Schwartz.[22] Let us consider the bilinear form (3.35) of Schwinger's variational principle for direct collisions in the case of scattering by a potential $V(\mathbf{r})$. Instead of concentrating on $\Psi(\mathbf{r})$ as the function to be varied, we define a new "trial function"

$$W(\mathbf{r}) \equiv V(\mathbf{r})\Psi(\mathbf{r}) \tag{3.53}$$

together with

$$\widetilde{W}(\mathbf{p}) = \int e^{-i\mathbf{p}\cdot\mathbf{r}} W(\mathbf{p}) d\mathbf{r} \tag{3.54}$$

Assuming that the Green's function is diagonal in momentum space, i.e.,

$$G_0^{(+)}(\mathbf{r}, \mathbf{r}') = (2\pi)^{-3} \int e^{i\mathbf{p}\cdot(\mathbf{r}-\mathbf{r}')} G_0^{(+)}(\mathbf{p}) d\mathbf{p} \tag{3.55}$$

we can rewrite the Schwinger form (3.35) as

$$[T_{if}] = \widetilde{W}^{(-)*}(\mathbf{k}_i) + \widetilde{W}_i^{(+)}(\mathbf{k}_f) - \int W_f^{(-)*}(\mathbf{r}) V^{-1}(\mathbf{r}) W_i^{(+)}(\mathbf{r}) d\mathbf{r}$$

$$+ (2\pi)^{-3} \int \widetilde{W}_f^{(-)*}(\mathbf{p}) G_0^{(+)}(\mathbf{p}) \widetilde{W}_i^{(+)}(\mathbf{p}) d\mathbf{p} \tag{3.56}$$

Thus, by using equation (3.57) we have to evaluate only single integrals, one in $\mathbf{r}$ space and one in $\mathbf{p}$ space, instead of the double integral

$$\iint \Psi_f^{(-)*}(\mathbf{r}') V(\mathbf{r}') G_0^{(+)}(\mathbf{r}, \mathbf{r}') V(\mathbf{r}) \Psi_i^{(+)}(\mathbf{r}) d\mathbf{r} \, d\mathbf{r}' \tag{3.57}$$

appearing in (3.35). The trial function $\widetilde{W}_i(\mathbf{p})$ is clearly the $T$ matrix $T_{ip}$ off the momentum shell. Thus, we can rewrite Schwinger's variational principle (3.35) as

$$[T] = 2T - TV^{-1}T + TGT \tag{3.58}$$

with the requirement that the second term on the right-hand side of (3.58) be evaluated in r-space and the third term in p-space. This slightly modified form of the Schwinger variational principle has been used successfully by Schwartz and Zemach[23] to solve a Bethe-Salpeter equation describing the scattering of two spinless mesons of equal mass, in the ladder approximation, and for energies below the inelastic threshold. The method, however, seems difficult to extend to the scattering by compound systems for which the Green's function has a complicated structure even in momentum space.

### 3.3. Schwinger Variational Principles for the Phase Shifts

The variational expression (3.36) provides also a direct way of obtaining variational principles for the phase shifts. If we develop the trial functions $\Psi_i^{(+)}$ and $\Psi_f^{(-)}$ in spherical harmonics and consider the coefficients as variational parameters, we obtain the following stationary expression

$$\left[\frac{\tan \eta_l}{k}\right] = \frac{\left[\int_0^\infty j_l(kr)U(r)u_l(r)r\,dr\right]^2}{\int_0^\infty dr \int_0^\infty dr'\, U(r)u_l(r)G_l(r,r')U(r')u_l(r')rr' - \int_0^\infty dr\, U(r)u_l^2(r)}$$

(3.59)

where

$$G_l(r,r') = kj_l(kr_<)n_l(kr_>)$$  (3.60)

while $j_l$ and $n_l$ are, respectively, the spherical Bessel and Neumann functions. Of course, one can directly establish this last variational principle by starting from the integral equation satisfied by the function $\bar{u}_l(r)$, namely,

$$\bar{u}_l(r) = rj_l(kr) + r\int_0^\infty G_l(r,r')U(r')\bar{u}_l(r')r'\,dr'$$  (3.61)

and using the relation

$$\tan \bar{\eta}_l = -k\int_0^\infty j_l(kr)U(r)\bar{u}_l(r)r\,dr$$  (3.62)

This last treatment[3-6] is in fact prior to the variational principle (3.36) for the transition amplitude. Its importance was crucial in nuclear theory since it allowed the first rigorous derivation of the effective range formulas[6] used in the analysis of low-energy nucleon–nucleon scattering phenomena.

In analogy with the Schwinger variational principle for the scattering amplitude we note that by taking

$$u_l(r) = rj_l(kr)$$  (3.63)

and replacing in the stationary expression (3.59) we get

$$\left[\frac{\tan \eta_l}{k}\right] = -\left[\int_0^\infty j_l^2(kr)U(r)r^2 dr\right.$$

$$\left. + \int_0^\infty dr \int_0^\infty dr'\, U(r)j_l(kr)G_l(r,r')U(r')j_l(r')r^2 r'^2 + \cdots\right]$$  (3.64)

which agrees with the Born series through the second order. Again, the expression which rests on a variational principle should be more accurate than the second Born approximation. We see also that with the approximation (3.63) the Hulthén–Kohn expression (2.22) for tan $\eta_l$ reduces again to the first Born approximation. Thus, Schwinger variational principles appear to be intrinsically superior to the corresponding variational principles based on differential equations, i.e., a good trial wave function should give better results when used in the Schwinger variational principle. However, this advantage is usually offset by the fact that the matrix elements involving Green's functions which appear in Schwinger's method are extremely tedious to evaluate, except in very simple cases. The much greater simplicity of the Hulthén–Kohn forms, on the other hand, allows for the use of more elaborate trial functions, even for relatively complicated problems.

## 4. MINIMUM PRINCIPLES IN SCATTERING THEORY

The scattering variational principles which we have developed thus far are stationary principles but they are not, in general, minimum (or maximum) principles. This fact constitutes an important difference compared to bound state problems, where upper bounds on the energy eigenvalues are readily obtainable. Thus, in scattering problems the variational result can fall either above or below the true value. Without a minimum (maximum) principle, the variational determination of the parameters is only the most plausible one (in the absence of other information) and greater flexibility in the choice of trial functions may even lead to less satisfactory results. Therefore, in applying the variational principles discussed in the two preceeding sections, great care must be taken and extensive work has is general to be done in order to obtain reliable results.

It is thus of considerable interest to develop variational principles giving upper and (or) lower bounds for phase shifts, scattering lengths, scattering amplitudes, etc. This was initiated by Kato[24,25] and developed extensively by Spruch *et al.*[26] As an introduction to these techniques, we shall derive in this section the Kato identity,[25] and then limit our discussion to the case of variational bounds for scattering lengths.

## 4.1. The Kato Identity

Let us start from equation (2.4) and adopt for $\bar{u}_l$ the "normalization"

$$\bar{u}_l(r) \xrightarrow[r\to\infty]{} \cos\left(kr - \frac{l\pi}{2} + \theta\right) + \cos(\bar{\eta}_l - \theta)\sin\left(kr - \frac{l\pi}{2} + \theta\right)$$

(4.1)

where $\theta$ is a fixed constant such that $0 \le \theta < \pi$. The trial function $u_l(r)$ has the same asymptotic behavior with $\bar{\eta}_l$ replaced by $\eta_l$

Now, let us apply Green's theorem

$$\int\limits_0^\infty \{fL_l[g] - gL_l[f]\}dr = \left[ f\frac{dg}{dr} - g\frac{df}{dr} \right]_0^\infty$$

(4.2)

where $f = \bar{u}_l$ and $g = u_l$. Recalling that $\bar{u}_l(0) = u_l(0) = 0$ and taking into account equation (4.1), we get

$$\int\limits_0^\infty \bar{u}_l L_l[u_l]\,dr = k\cot(\eta_l - \theta) - k\cot(\bar{\eta}_l - \theta)$$

(4.3)

Setting

$$w_l(r) = u_l(r) - \bar{u}_l(r)$$

(4.4)

and noting that

$$L_l[w_l] = L_l[u_l]$$

(4.5)

we obtain the Kato identity[25]

$$k\cot(\bar{\eta}_l - \theta) = k\cot(\eta_l - \theta)$$

$$-\int\limits_0^\infty u_l L_l[u_l]dr + \int\limits_0^\infty W_l L_l[w_l]dr$$

(4.6)

The first two terms on the right of equation (4.6) can be calculated by specifying the trial function $u_l$ and give an approximate value of the quantity $k\cot(\bar{\eta} - \theta)$, while the last term (of order $w_l^2$) should be small if $u_l$ is a good approximation to $\bar{u}_l$. In fact, for the choice $\theta = \pi/2$, the neglect of the last term on the right side of equation

(4.6) leads to

$$[k \tan \eta_l] = k \tan \eta_l + \int_0^\infty u_l L_l[u_l]dr \qquad (4.7)$$

which is precisely the Hulthén–Kohn variational principle (2.22). The above discussion can easily be generalized to elastic scattering by compound systems. A generalized version of the Kato identity can also be obtained for multichannel scattering.[26]

Since equation (4.6) is an exact formula, one should be able to obtain bounds on $k \cot(\bar\eta_l - \theta)$ and, hence, on $\bar\eta_l$, if the last term on the right of equation (4.6) can be estimated. We illustrate this point in the next paragraph by showing how variational bounds can be obtained for scattering lengths.

## 4.2. Variational Bounds for Scattering Lengths

Let us begin by considering $S$-wave scattering by a short-range potential at zero incident energy.[27] Choosing $\theta = \pi/2$ in equation (4.6), we have (with $L_0 \equiv L$, $u_0 \equiv u$)

$$\bar a = a - \int_0^\infty u L[u]dr \qquad (4.8)$$

where $\bar a$ and $a$ denote, respectively, the exact and trial scattering lengths and

$$u(0) = \bar u(0) = 0 \qquad (4.9)$$

$$\bar u(r) \xrightarrow[r\to\infty]{} -r + \bar a \qquad (4.10)$$

and

$$u(r) \xrightarrow[r\to\infty]{} -r + a \qquad (4.11)$$

We note that the function $w = u - \bar u$ is such that

$$w(0) = 0 \qquad (4.12)$$

and

$$w(r) \xrightarrow[r\to\infty]{} a - \bar a \qquad (4.13)$$

Now, if the potential cannot support a bound state, one has

$$\int_0^\infty wL[w]dr \le 0 \qquad (4.14)$$

To prove this statement, we first note that if no bound state exists, $L$ is a negative definite operator on the space of quadratically integrable functions. Now equation (4.13) shows that the function $w(r)$ approaches a constant value for large $r$ and is therefore not quadratically integrable. If, however, we introduce

$$w_\lambda(r) \equiv w(r)e^{-\lambda r} \tag{4.15}$$

with $\lambda > 0$, we have

$$M(\lambda) \equiv \int_0^\infty w_\lambda(r)L[w_\lambda(r)]dr \leq 0 \tag{4.16}$$

To show that $M(0)$ is not positive, it remains to prove that $M(\lambda)$ is continuous at $\lambda = 0$. This is easily done by considering the quantity

$$M(\lambda) - M(0) = \int_0^\infty w(r)[e^{-2\lambda r} - 1]L[w(r)]dr$$

$$+ \lambda^2 \int_0^\infty w^2(r)e^{-2\lambda r}\,dr - 2\lambda \int_0^\infty w(r)w'(r)e^{-2\lambda r}\,dr \tag{4.17}$$

and checking that each of the three terms on the right of equation (4.17) vanishes in the limit $\lambda \to 0$. Thus, if no bound state exists, the scattering length satisfies the relation

$$\bar{a} \leq [a] = a - \int_0^\infty uL[u]dr \tag{4.18}$$

with

$$u(0) = 0 \tag{4.19}$$

and

$$u(r) \xrightarrow[r \to \infty]{} - r + a \tag{4.20}$$

We recognize on the right-hand side of equation (4.18) the Hulthén–Kohn variational principle for the scattering length, which, as we see, gives a bound on $\bar{a}$ under the above specified conditions. The procedure to determine the variational parameters is now perfectly unambiguous. For a trial function $u(c_1, c_2, \ldots, c_n; a; r)$ one should set

$$\frac{\partial[a]}{\partial c_i} = 0 \qquad (i = 1, 2, \ldots, n) \tag{4.21}$$

and

$$\frac{\partial[a]}{\partial a} = 0 \tag{4.22}$$

For linear variational parameters, equation (4.21) and (4.22) lead to a unique set of values for the parameters. For nonlinear variational parameters, there in general more than one set of parameters, and therefore equation (4.18) tells us that one shoud choose the set which gives the lowest value of $[a]$. We note, incidentally, that in the type of problem considered here, this procedure [referred to as the Hulthén-Kohn method in paragraph (2.1)] is definitely superior to the original Hulthén prescription, which in this case can be written

$$\frac{\partial[a]}{\partial c_i} = 0 \tag{4.23}$$

and

$$\int_0^\infty uL[u] = 0 \tag{4.24}$$

This follows from the fact that for a given form of the trial function the variational parameters evaluated according to equation (4.21) and (4.22) give the smallest value for $[a]$, thus providing the best approximation to the correct scattering length.

The preceding considerations can easily be generalized to situations for which $l \neq 0$,[27] to the case of a potential which is the sum of a short-range potential and a repulsive potential, and for potentials which vanish as $r^{-n}$ for $r \to \infty$.[28] The method can also be extended to the scattering of one compound system by another with zero initial relative kinetic energy.[26]

Generalizations of the method to determine bounds on scattering lengths when bound states are present have been given[29] for arbitrary $l$-values, including Coulomb and other long-range potentials, and scattering of one compound system by another. Applications of these techniques have been made by Spruch et al. to various problems such as the determination of neutron-deuteron doublet and quartet scatterering lengths[30,31] and the zero energy scattering of positrons[32] and electrons[33] by hydrogen atoms. For these last problems, Schwartz[15] has obtained the very accurate results (in units of

Bohr radius)

$$a_s = 5.965 \pm 0.003$$
$$a_t = 1.7686 \pm 0.0002$$

and

$$a_+ \leq -2.10$$

where $a_s$ and $a_t$ are, respectively, the singlet and triplet electron-hydrogen scattering lengths and $a_+$ denotes the positron-hydrogen scattering length.

Further generalizations of variational bounds for non-zero energies and multi-channel scattering have been proposed,[34,35] but on theoretical as well as practical grounds the situation is less satisfactory, due to the very complicated nature of the problem. If is hoped that eventually these methods will provide a reliable calculational scheme for handling scattering problems within the framework of a given dynamical theory.

## REFERENCES

1. L. Hulthén, *Kgl. Fysiogr. Sallsk Lund Forh*, **14**(21): (1944).
2. I. G. Tamm, *J. Exp. Theor. Phys. USSR.*, **14**: 21 (1944).
3. J. Schwinger, unpublished lectures (1947); *Phys. Rev.*, **72**: 742 (1947).
4. L. Hulthen, *Ark. Math. Ast. Phys.* **35A** (25): (1948).
5. W. Kohn, *Phys. Rev.* **74**: 1763 (1948).
6. J. M. Blatt and J. D. Jackson, *Phys. Rev.*, **76**: 18 (1949).
7. B. A. Lippmann and J. Schwinger, *Phys. Rev.* **79**: 469 (1950).
8. B. A. Lippmann, *Phys. Rev.* **102**: 264 (1956).
9. C. Joachain, *Nucl. Phys.* **64**: 548 (1965)·
10. See, for example, H. S. W. Massey's article in "Handbuch der Physik," ed S. Flugge, **36**: 285 (1956).
11. H. S. W. Massey and B. L. Moiseiwitsch, Proc. Roy. Soc. London **A205**: 483 (1950).
12. B. L. Moiseiwitsch, *Proc. Roy. Soc., London*, **A219**: 102 (1953).
13. S. S. Huang, *Phys. Rev.*, **76**: 477, 1878 (1949).
14. C. Schwartz, *Ann. Phys.* (New York), **16**: 36 (1961).
15. C. Schwartz, Phys. Rev., **124**: 1468 (1961).
16. M. L. Goldberger and K. M. Watson, "Collision Theory," John Wiley, Inc., New York, 1964, p. 85.
17. E. Gerjuoy and D. S. Saxon, *Phys. Rev.*, **94**: 484 (1954).
18. C. Joachain, *Nucl. Phys.* **64**: 529 (1965).
19. C. Joachain, *Ann. Inst. Henri Poincaré*, **1**: 55 (1964).

20. C. Joachain, Bull. *Cl. Sc. Acad. Roy. Belg.*, **48**: 302 (1962).
21. Y. Yama uchi, *Phys. Rev.*, **95**: 1628 (1954).
22. C. Schwartz, *Phys. Rev.* **141** (4): 1468 (1966).
23. C. Schwartz, and C. Zemach, *Phys. Rev.* **141** (4): 1454 (1966).
24. T. Kato, *Progr. Theor. Phys.*, **6**: 295 (1951).
25. T. Kato, *Progr. Theor. Phys.* **6**: 394 (1951).
26. L. Spruch, "Minimum Principles in Scattering Theory," *Lectures in Theoretical Physics, Vol. IV*, Interscience Publ. 1962, p. 161.
27. L. Spruch and L. Rosenberg, *Phys. Rev.*, **116**: 1034 (1959).
28. T. F. O'Malley, L. Spruch, and L. Rosenberg, *J. Math. Phys.* **2**: 491 (1961).
29. L. Rosenberg, L. Spruch, and T. O'Malley, *Phys. Rev.*, **118**: 184 (1960).
30. L. Spruch and L. Rosenberg, *Phys. Rev.* **117**: 1095 (1960).
31. L. Spruch and L. Rosenberg, *Nucl. Phys.* **17**: 30 (1960).
32. L. Spruch and L. Rosenberg, *Phys. Rev.* **117**: 143 (1960).
33. L. Rosenberg, L. Spruch and T. F. O'Malley, *Phys. Rev.*, **119**: 164 (1960).
34. See, for example, Y. Hahn, and L. Spruch, *phys. Rev.* **153** (4): 1159 (1967); L. Rosenberg, *Phys. Rev.* **B138**: 1343 (1965).
35. R. Sugar and R. Blankenbecler, *Phys. Rev.*, **B136**: 472 (1964).

# The $(\pi^+, 2p)$ Reactions on Nuclei

T. Bressani, G. Charpak, J. Favier,
L. Massonnet,‡ W. E. Meyerhof§ and
Č. Zupančič

*CERN*
*Geneva, Switzerland*

## 1. INTRODUCTION

When a slow pion is absorbed in a nucleus by a single nucleon which takes all the available energy, the conservation of energy and momentum requires the nucleon to have a large momentum inside the nucleus before the absorption process. For a pion at rest this momentum would be 500 MeV/c, and it is known that not many nucleons have such high momenta. In the case of an absorption by a pair of nucleons, on the other hand, energy and momenta are easily balanced, the two nucleons going in opposite directions with a high relative momentum and a small momentum of their center of mass. The fact that the absorption by a pair of nucleons is favored makes pion absorption an ideal tool for the study of residual states with two holes in the internal shells.

In the impulse approximation, the nucleons not participating in the reaction are just spectators, and their momentum in the laboratory is the momentum they had relative to the pair of nucleons absorbing the pion, at the instant of the absorption. If well separat-

‡Present address: Institut de Physique Nucléaire, Orsay, France.
§U.S. National Science Foundation Senior Postdoctoral Fellow, on leave from Stanford University, California, USA, 1966-7.

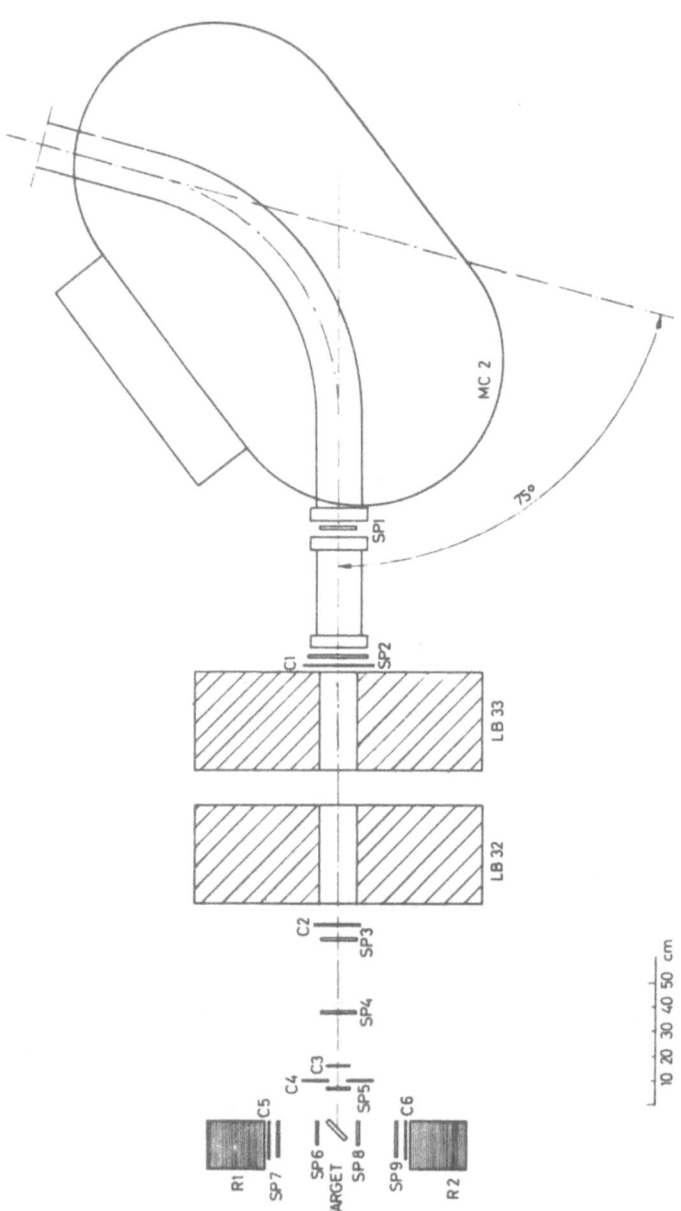

Fig. 1. Experimental set-up of the $(\pi^+, 2p)$ experiment at 80 MeV.

ed final excited states can be obtained from the $(\pi^+, 2p)$ reactions, a study of the momentum distribution in the laboratory of these nuclei gives the distribution of momentum of these configurations relative to the pair of absorbing nucleons. The choice of the geometry of detection can strongly influence the region of momentum distributions contributing to the reaction. We shall discuss this point when we come to the results obtained with $^6$Li. Not very much theoretical work has been done since the first papers underlining the general interest of these reactions,[1,7] and only the very recent work of Kopaleishvili[4,5] enters into the detailed structure of some specific nuclei. We shall summarize their results when discussing the results obtained for $^{12}$C, $^{14}$N, and $^{16}$O. In this report we wish to give a review of the results, and to show how necessary it is to have more theoretical investigations of this field.

## 2. EXPERIMENTAL TECHNIQUES

The 80-MeV $\pi^+$ beam from the CERN Synchro-Cyclotron is impinging on a target T (Fig. 1). The beam is bent through 75° by the magnet MC2, and two spark chambers (SP1 and SP2) at the exit of MC2 determine the direction and the position of the pions. Since the input direction of the pion beam, as determined by the collimation through the beam-transport system, is uncertain at most by $\pm 0.5°$, this allows a momentum determination with an error of the order of 1%. The beam-energy spread is 4 MeV (FWHM). We did not use spark chambers in front of the bending magnet since the flux of particles coming straight through the pipe gives a high probability of having two sparks within the memory of a spark chamber ($\sim 1\mu$ sec). Using current-division filmless chambers[8] which do not permit the detection of a double spark, we have chosen to bend the beam through a larger angle so that the uncertainties of the initial input position would be of less importance. This results in a beam which is optically worse after such a bending. A quadrupole pair focuses the pion beam onto the target T after passage through three spark chambers (SP3, SP4, SP5) giving, with redundancy, the line of flight of the pions impinging upon the target. Each of the two protons, emitted at $(79 \pm 10)°$ from the line of flight of the pions, is detected in two spark chambers (SP6, SP7, SP8, SP9), giving the spatial coordinates. It is stopped in a range assembly (R1, R2) consisting of 50 scintillators viewed by 50 photomultipliers. The choice of this particular geometry was dictated by the fact that it corresponds, in the case of a $^6$Li target, with protons of equal energy, to a zero recoil energy for the residual helium in the ground state. The protons also traverse two scintillation counters of 8 mm thickness ($C_5$ and $C_6$). An event is defined by a coincidence $C_1 C_2 C_3 \bar{C}_4 C_5 C_6$; $\bar{C}_4$ is a counter with a hole of $4 \times 4$ cm, vetoing the incoming beam, and defining the active area of the target. Typically, 25.000 $\pi^+$ of 80 MeV, with a duty cycle of about 30 to 50%, hit the target per second. About 50% of the pion intensity is rejected by the veto counter defining the beam.

All the spark chambers are automatic and for each event we collect the following data:

18 coordinates from the 9 spark chambers SP1, 2, 3, 4, 5, 6, 7, 8, 9.

Two pulse heights from the linear outputs of the counters $C_5$ and $C_6$. This allows a separation between protons and pions.

The pattern of the activated counters in the range assemblies $R_1$ and $R_2$.

A logic signal indicating whether within the memory time of the spark chambers (1 $\mu$sec) more than one particle went through. This is done simply by counting the number of particles which traversed the beam counter during the memory time.

The digitized data are stored on an IBM magnetic-tape unit, but are first processed by a PDP 8. We found that it was an essential advantage to have a computer on line. The complexity of the experiment makes it very difficult to check every parameter with sufficient regularity. The computer on line saves a great deal of machine time by giving a constant check on the parameters of the experiment.

## 3. EXPERIMENTAL RESULTS

The data stored on the magnetic tape are analyzed by a CDC 6600. In the reaction

$$\pi^+ + A \rightarrow p + p + B$$

we calculate, among other quantities, the excitation energy of $B$, the three components of the recoil momentum of $B$ and the total recoil momentum of $B$.

The following partial results have been obtained with a series of nuclei. We wish to emphasize that the data are not definitive. For instance, uncertainties of 2 MeV or even 4 MeV can exist for the origin of excitation energies. A careful adjustment of this point is now under way.

## I. $^6$Li

This is the target that was most studied, and that dictated our choice of geometry. For a pion of 80 MeV being absorbed by $^6$Li and leading to a reaction where the residual $^4$He nucleus is in its ground state, with the two protons of equal energy coplanar with the pion, this leads to a zero momentum for the recoil. If we take a model in which the absorption occurs in a pair ($np$) while the helium core is merely a spectator, in the impulse approximation, the momentum $K$ of the recoiling helium is given by $K = -(k_1 + k_2 - q)$, where $k_1$ and $k_2$ are the momenta

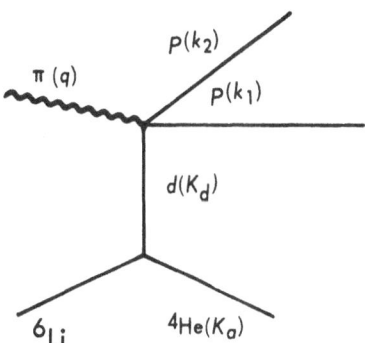

of the protons, and $q$ is the momentum of the incident pion; $K = -K_d$, the momentum of the deuterium inside the $^6$Li nucleus. Perturbation theory in its simplest form gives the result that the reaction is proportional to

$$g^2(K_d^2)f^2(\Delta^2)$$

where $g^2$ is the probability of finding a deuteron of momentum $K_d$, and $f^2$ is the probability of finding, inside the cluster, two nucleons with relative momentum $\Delta$. It is assumed that these functions are slowly varying functions when their arguments are large, while they have a maximum when the argument is zero. By choosing geometrical conditions where one of the arguments is zero, it is expected that the study of the variations of the cross section around this geometry will give information about this function. These considerations of Jean[1] dictated our choice of geometry. However, the fact that we had to choose a large solid angle for the protons and a wide energy acceptance gives us a wide acceptance in the recoil momentum spectrum. From phase-space considerations alone, the density around zero momentum is multiplied by $p^2$ so that we have, in fact, no events with zero recoil energy.

Figure 2 show the spectrum of the excitation energy of the residual nucleus.

Figure 3 gives a two-dimensional plot of the recoil momentum *vs.* excitation energy.

Figures 4a, b, c, d, e show the variation of the excitation energy for the various bands of the recoil momenta.

We see that the transition to the ground state corresponds to a peripheral process with a small momentum transfer. It is possible to calculate the momentum distribution assuming a simple cluster model, with a Hulthén-type wave function describing the relative motion of the deuteron and the helium core. Figure 5a shows the

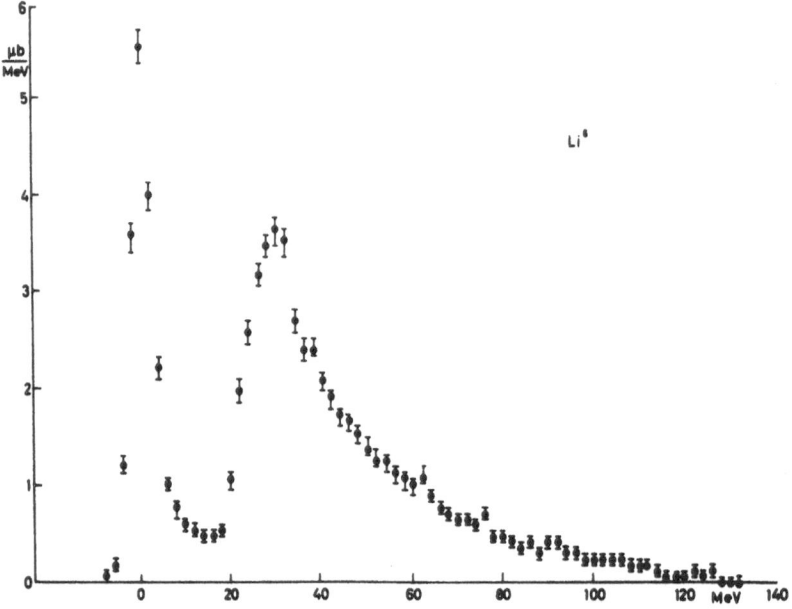

Fig. 2. Excitation-energy distribution of the recoil $^4$He. Reaction $\pi^+ + {}^6\text{Li} \rightarrow p + p + {}^4\text{He}$.

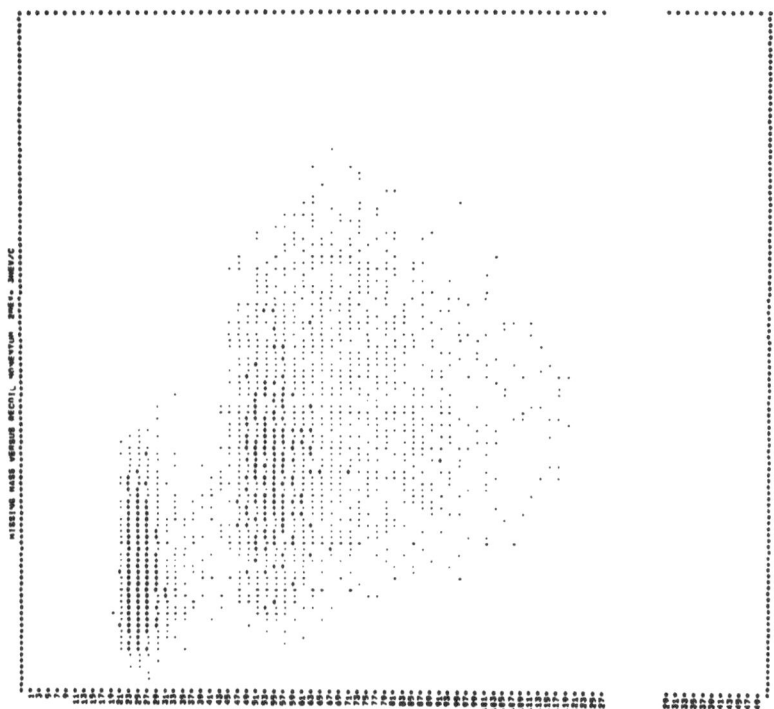

Fig. 3. Two-dimensional plot of the recoil momentum *vs.* excitation energy for ⁶Li.

blank : 0, 1, 2 counts
·   : 3 to 5 counts
+   : 9 to 14 counts
×   : above 14 counts.

The zero on the abscissa is at the point labeled 25, and the spacing is 2MeV; the points on the ordinate are spaced by 3MeV/c.

experimental momentum distribution of the ⁴He recoils produced in the transitions to the ground state, together with a Monte-Carlo calculation of the distribution expected under our experimental conditions for such a model.[3,9]

Figure 5b shows the distribution of the recoil momentum with respect to the pion beam. The distribution is symmetrical in the forward-backward directions in the laboratory showing the validity of the impulse approximation in the treatment of this problem.

We see from Fig. 4 that for recoil momenta up to 150 MeV, the second peak is mainly at 30 MeV. From 150 MeV/c onwards, the transition leads also to states close to 50 MeV. Some light has been thrown on the peak at 30 MeV by the study of ⁴He.

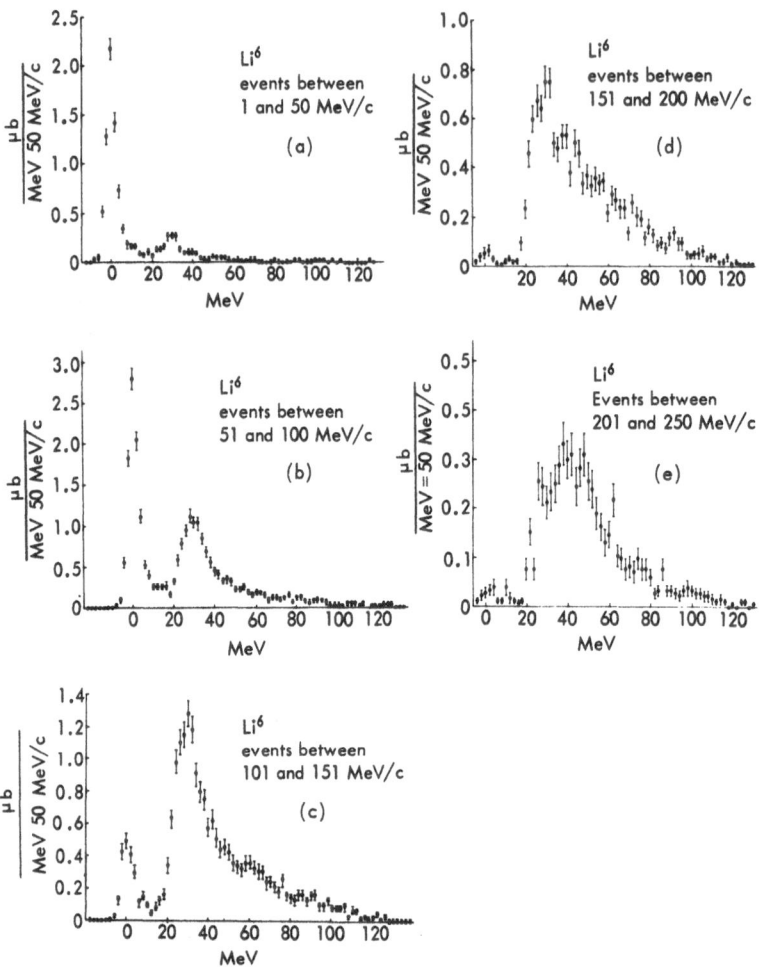

Fig. 4. Excitation energy for different bands of recoil momenta; ⁶Li target a) 1 to 51 MeV/c; b) 51 to 101 MeV/c; c) 101 to 151 MeV/c; d) 151 to 201 MeV/; e) 201 to 251 MeV/c.

## II. ⁴He

We used a liquid-helium target consisting of a vertical cylinder of 5 cm radius. Figure 6 shows the excitation energy of the residual pair of nucleons. We observe that a large fraction of the $(\pi^+, 2p)$ reactions leave the $np$ pair in a strongly interacting state. The width of the peak, 13 MeV, is larger than our experimental resolution (6 MeV), but narrower than the 30-MeV peak in the $^6$Li. This peak can be explained within the framework of the cluster model by assuming that it is due to the absorption of the pion in the helium core. Figure 7 shows a Monte Carlo calculation of the excitation energy of the recoil from a $(\pi^+, 2p)$ reaction on $^6$Li, assuming that the absorption occurs in the helium.[3] For this calculation a naive bi-deuteron model is taken for the helium, in the absence, for the time being, of any realistic calculation fitting the helium results. Figures 4b and 4c show that the peak structure of the $np$ recoil excitation is maintained even for the highest recoil momenta as is also observed for the 30 MeV peak in $^6$Li. The comparison of the cross section of the $(\pi^+, 2p)$ reaction in D, $^4$He, and $^6$Li is also of importance when checking these considerations on the $^6$Li model:

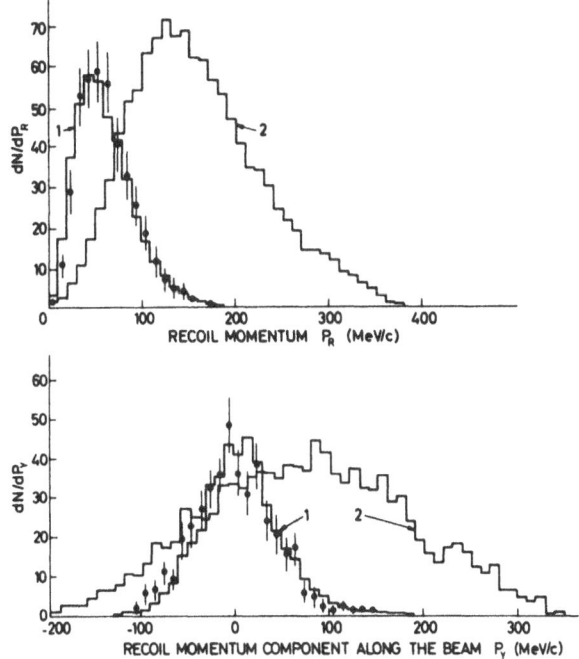

Fig. 5. a) Recoil momentum spectrum for the reactions leading to the ground state of $^4$He. Histogram 1 is the result of a Monte-Carlo calculation assuming a peripheral mechanism on the deuterium cluster in $^6$Li. Histogram 2 is the distribution expected from a phase-space calculation. The points are our experimental results.

b) The same for the projection of recoil momentum along the beam axis.

we shall come back to this below, when we discuss the general question of cross sections.

## III. $^7$Li

Figure 8 shows the excitation energy of the $^5$He recoil. It again seems that a simple cluster model explains the two-peak structure. The second peak is distant by about 22 MeV from the ground state.

## IV. $^{12}$C, $^{14}$N, $^{16}$O

These three nuclei have been studied with targets of graphite, liquid nitrogen, and water.[5] We group them together since Kopalei-shvili et al.[6] have made definite predictions concerning the proba-bility of exciting the various levels of the residual nuclei.

The calculations are made in the independent-pair model for the target nuclei. The interaction of the ejected nucleons with the residual nucleus is neglected. The interaction between the ejected

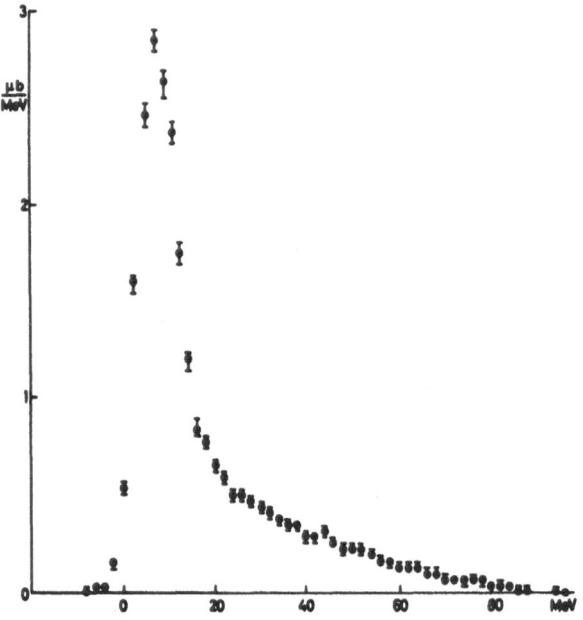

Fig. 6a. Excitation energy of the recoiling $np$ pair. Reaction $\pi^+ + \,^4\mathrm{He} \rightarrow p + p + (np)$; Excitation energy for all recoil momenta.

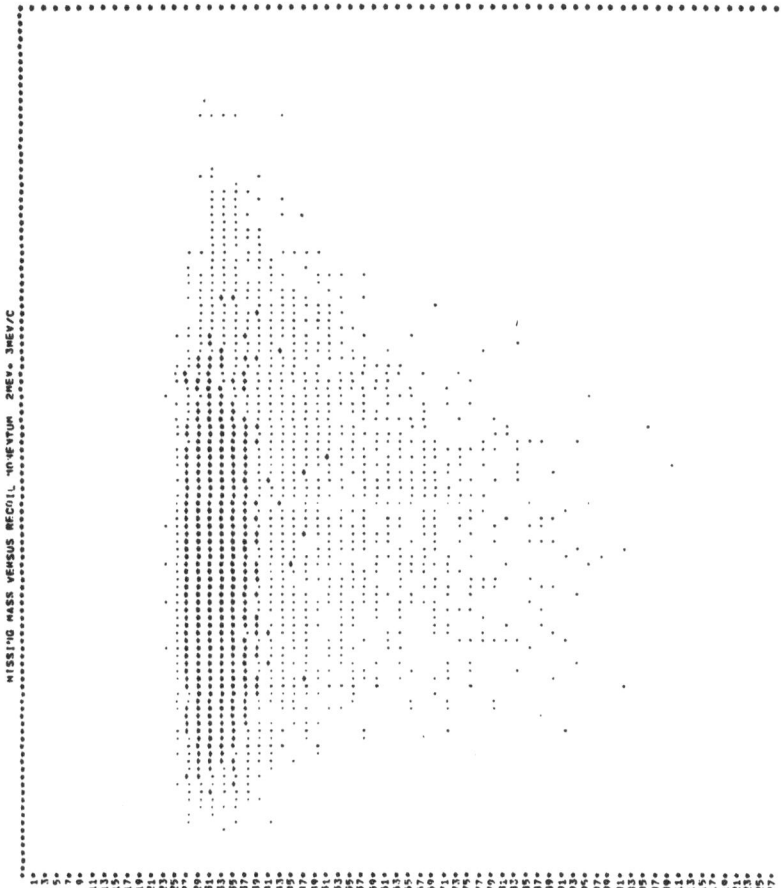

Fig. 6b. Excitation energy of the recoiling $np$ pair. Reaction $\pi^+ + {}^4\mathrm{He} \rightarrow p + p +$ $(np)$; Two-dimensional plot of the recoil momentum $\bar{v}s$. excitation energy. The scales on the abscissa and on the ordinate are the same as in Fig. 3.

nucleons is taken into account in the asymptotic approximation. The calculations are made for $E_\pi = 40\,\mathrm{MeV}$, since experimental data are available for the phases of $N$–$N$ scattering at the corresponding energies. They find that the main contribution to the absorption comes from the $p$-shell. The probability for the absorption by a nucleon in both the $S$ and the $P$ shell is estimated to be at most 15%. The wave function of the nucleons of the $P$-shell is formed from all possible products of the wave functions of the two nucleons absorbing a pion and $(n\text{-}2)$ other nucleons. With this model they take into account all the levels of the residual nuclei with $E_K \lesssim 20\,\mathrm{MeV}$. Although they made the calculation with $40\,\mathrm{MeV}$

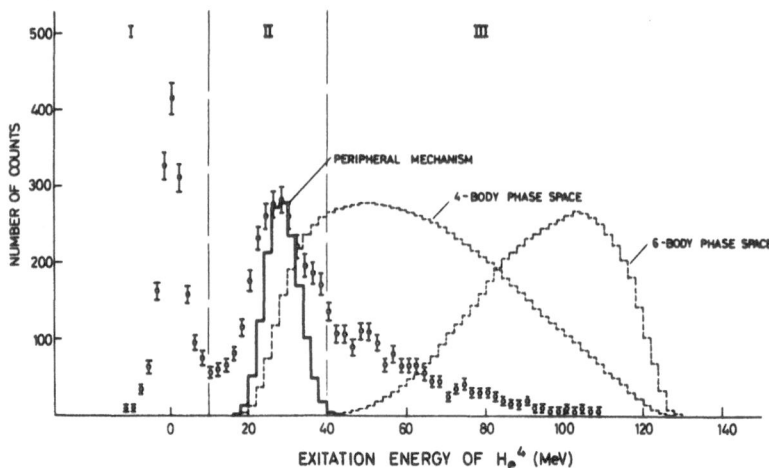

Fig. 7. Excitation energy of the recoil in the reaction $\pi^+ + {}^6Li \rightarrow p + p + {}^4He$. Histogram: Result of a Monte-Carlo calculation assuming a peripheral mechanism for the absorption in the ${}^4He$ core of ${}^6Li$.

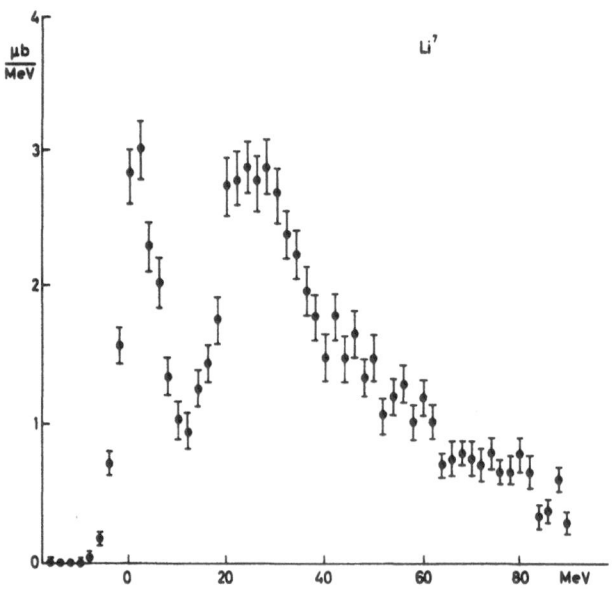

Fig. 8. Excitation energy of the recoil in the reaction $\pi^+ + {}^7Li \rightarrow p + p + {}^5He$.

pions against 80 MeV in our case, and for a different geometry, it seemed to us interesting to fold our experimental resolution into their calculations and compare them with our results. These are the solid curves traced on Fig. 9. The cross section scale has been adjusted arbitrarily, but we see that these authors make realistic predictions concerning the density of excitation of the different states.

Clearly, such calculations should be extended to cover the range of nuclei that we have studied, and the energy of pions we have

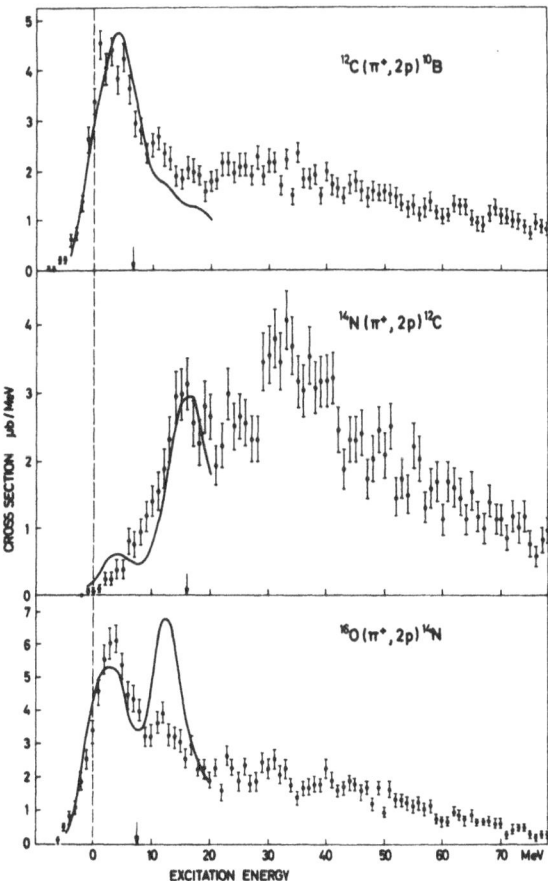

Fig. 9. Excitation energy of the recoils with targets of $^{12}$C, $^{14}$N, and $^{16}$O. Solid curves: predictions of Kopaleishvili *et al.*[4][5] with our energy resolution. Vertical scale arbitrary.

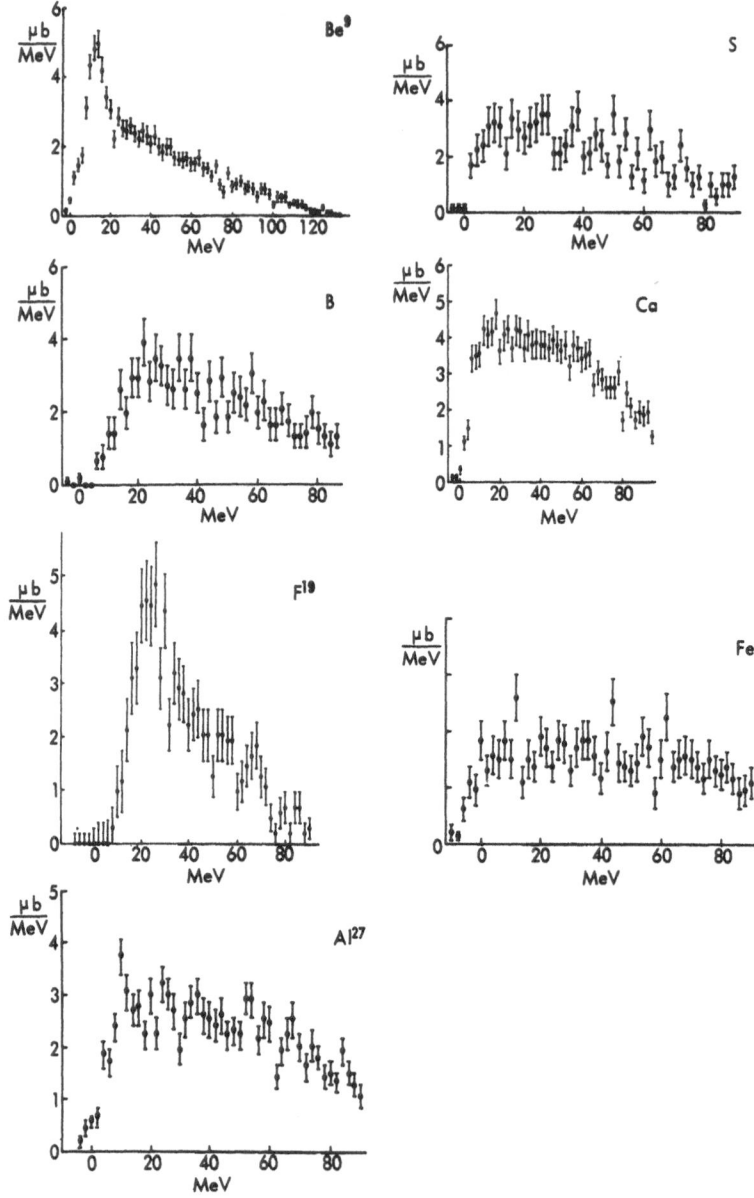

Fig. 10. Excitation energy of the recoils with targets of $^9$Be, B, $^{19}$F, $^{27}$Al, S, $^{40}$Ca, Fe.

worked with. It is striking that strong differences appear in the energy spectrum above 20 MeV, between $^{12}$C and $^{14}$N. The message that is contained here concerning the properties of different nuclear structures is apparently out of reach of the theory for the moment.

## V. $^9$Be, B, $^{19}$F, $^{27}$Al, S, Cl, $^{40}$Ca, Fe, and Pb

For these nuclei let us display the spectrum of the excitation energy (Fig. 10). As expected, all structures are smoothed out with increasing atomic number.

## 4. CROSS SECTIONS

Since our experiment has been done in a fixed geometry subtending a small fraction of the $4\pi$ solid angle (5% for each proton), we cannot extract much information from the measurement of the cross sections. It is known from the measurements of the

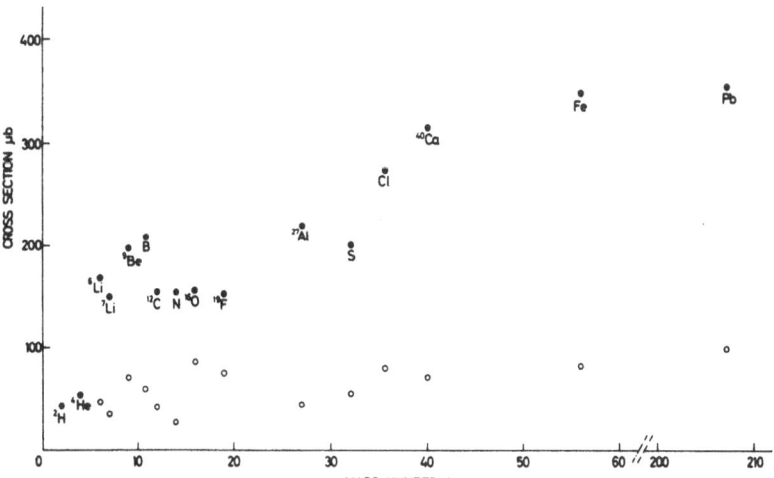

Fig. 11. Differential cross section integrated over our solid angle and energy acceptance.

$$\sigma = \int_{\text{instrument}} \frac{d^4\sigma}{dE_1 dE_2 d\Omega_1 d\Omega_2}$$

The solid angle subtended by a proton detector is 5% of $4\pi$. Solid circles: cross section integrated over the entire excitation energy range measured; open circles: cross section integrated up to 20 MeV excitation energy of the residual nucleus.

angular distributions of the protons in the $(\pi^+, 2p)$ reactions in deuterium[10,11] that, in this case, the angular distribution is strongly anisotropic, of the form $1 + 3.7\cos^2$ at 80 MeV. We are sitting at the position of the minimum cross section for deuterium. It is likely that these distributions vary for the reactions occurring on pairs sitting in different shells of the nuclei.

Thus, a correction factor varying for each energy band of the excitation energy would be necessary. The differential cross section integrated over the solid angle and energy band of our instrument is displayed on Fig. 11. We see, for instance, that the cross section of $^6$Li is much higher than the sum of cross sections of $^2$D and $^4$He (168 $\mu$b) against 43 $\mu$b and 53 $\mu$b for $^2$D and $^4$He. This shows that pairs of nucleons other than those in the deuterium and helium clusters play a role in the absorption. More interesting, perhaps, are the cross sections for the reactions leading to the ground states of the residual nuclei. Limiting ourselves to the first 20 MeV of excitation energy, we see strong variations displaying the differences between the nuclear species. We may notice that the cross section for the peripheral reaction on $^6$Li is the same as the cross section on deuterium.

## 5. THE $(\pi^+, 2p)$ REACTION AS A FUNCTION OF ENERGY

We have made a brief study of the reaction $(\pi^+, 2p)$ on $^6$Li at 40, 80, 150, 200, and 260 MeV. We observe a variation of the cross section as we go through the $(\frac{3}{2}, \frac{3}{2})$ resonance region, for the reactions leading to the ground state and the first excited state. Between 40 and 200 MeV, the spectrum of the excitation energy varies very little. At 260 MeV, the structure is more complex, and more experiments are necessary to understand the data.

## 6. CONCLUSION

In conclusion, we may say that we have obtained a set of data which is raw material for which more theoretical investigation is necessary. It appears to us that since definite nuclear-structure effects appear, as shown by the comparison of our results in $^{12}$C, $^{16}$O, and $^{14}$N with the calculations of Kopaleishvili, it will be necessary to undertake these experiments with more refined techniques giving a better resolution. The limitation in our energy resolution comes mainly from the range chambers. We have to measure two-proton spectra ranging from 40 MeV to 200 MeV. An accuracy of about 3 MeV with the range chambers is the best we have reached. The use of magnetic spectrometers and spark-chamber techniques to measure the energy of the charged particles would lead to 1-MeV resolution, as can easily be seen from the experience already acquired in this field by many groups. It would also be useful to cover a $4\pi$ angle with respect to the incoming pion.

## REFERENCES

1. M. Jean, Proc. Int. Symposium on Direct Interactions and Nuclear Reaction Mechanism, Padua (1962); *Nuovo Cim. Suppl. I*, **2**: 400 (1964).
2. G. Charpak, G. Gregoire, L. Massonnet, J. Saudinos, J. Favier, M. Gusakow and M. Jean, *Phy. Letters* **16**: 54 (1965).
3. G. Charpak, J. Favier, L. Massonnet and Č. Zupančič, Int. Conf. on Nuclear Physics, Gatlinburg, Tennessee, 12–17 Sept. 1966. CERN Preprint 66/1000/5.
4. T.I. Kopaleishvili, I.Z. Machabeli, G. Sh. Goksadze and N.B. Krupennikova, *Phy. Letters* **22**: 181 (1966).
5. T.I. Kopaleishvili, *Nucl. Phys.* **B1**: 335 (1967).
6. J. Favier, T. Bressani, G. Charpak, L. Massonnet, W.E. Meyerhof and Č. Zupančič, *Phys. Letters* **25B**: 409 (1967).
7. T. Ericson, *Phy. Letters* **2**: 278 (1962).
8. G. Charpak, J. Favier, and L. Massonnet, Int. Conf. on Instrumentation for High-Energy Physics, Stanford, Cal., 9–10 Sept. 1966. CERN Preprint 66/1267/5.
9. Č. Zupančič, in: "High-Energy Physics and Nuclear Structure," North-Holland, Amsterdam (1967), p. 171.
10. R. Durbin, H. Loar and J. Steinberger, *Phys. Rev.* **84**: 581 (1951).
11. B.S. Neganov and L.B. Parfenov, *Zh. Eksperim. i Teor. Fiz.* **34**: (1958), 767; [English transl. Soviet Physics JETP **7**: 528 (1958)].

# Muon Capture as a Probe for Nuclear Structure

## M. Rho

*C. E. N.*
*Saclay, France*

---

## 1. INTRODUCTION

The process of muon capture in complex nuclei $\mu^- + N_i \to \nu_\mu + N_f$ where $N_i$ and $N_f$ refer, respectively, to initial and final nuclei has so far been used as a means of studying the fundamental weak interaction. In particular considerable work has been done both theoretically and experimentally to check the "universal fermi interaction" hypothesis, and to determine the pseudoscalar coupling constant which becomes important in the $\mu$-capture due to a large momentum transfer involved in the process. However, except in very special cases, the nuclear structure complications introduce a large uncertainty in the analysis and hence make it very difficult to learn anything meaningful about the fundamental process. For the study of the fundamental weak interaction, therefore, one ought to resort to a process entirely free of nuclear structure, in particular $\mu^- + p \to \nu_\mu + n$. Many laboratories are actively engaged in this work. This is significant not only for weak interaction physics, but also for nuclear structure physics. One topic on just how the clarification of the weak process would help us in studying nuclear structure will be presented in this paper.

## 2. HYPOTHESES

For the time being, let us assume that the weak interaction mechanism is known. In particular the Conserved Vector Current

(CVC) and the Partially Conserved Axial-Vector Current (PCAC) hypotheses are taken to be valid. The CVC has by now been tested many times and can be considered to be established.[1] There is no direct check for the PCAC, and the second hypothesis seems to be completely open to question. Nevertheless for the muon-capture process, there does not exist a conclusive evidence that the PCAC (with the one-pion-pole dominance approximation) is inconsistent with experiments, although at this moment, the asymmetry parameter of the neutron angular distribution following the capture of muon[2] and the radiative muon capture[3] cannot be understood with the Goldberger–Treiman value of the pseudoscalar coupling constant. But it is my feeling that the discrepancy lies in the ignorance of the nuclear process in the first case and in the lack of knowledge about some of the (strong interaction) structure-dependent terms in the second case.[4]

In calculating the transition rates $\mu^- + N_i(J_i^\pi, T_i) \rightarrow \nu_\mu + N_f(J_f^\pi, T_f)$, one can invoke the weak interaction hypotheses in two different ways: 1. If the initial and final states can be considered as elementary particle states, then the hypotheses can be applied to the *nuclear* form factors. This is the approach of Kim and Primakoff, and is mostly applicable to bound-to-bound state transitions; 2. The hypotheses may be invoked in the *nucleon* form factors. In this case, the application to nuclei involves the impulse approximation and requires the knowledge of nuclear wave functions. The first approach tells us nothing about the nuclear structure, but as we will see shortly it provides a convenient means of relating different processes. If we can assess how good the neglect of exchange current contribution is, the second approach can be used with the help of the information given by the first method to gain insight into what goes on in nuclei.

## 3. MUON CAPTURE AND PHOTOPION PRODUCTION

Let us describe here the approach of Kim and Primakoff (called "elementary particle method" or in short EPM.[5] For definiteness consider the process

$$\mu^- + O^{16}(0^+) \rightarrow \nu_\mu + N^{16}(2^-, T = 1) \tag{1}$$

The matrix element of the hadronic current appropriate for this process is to order $E_\nu/2m_p$

$$\langle 2^- |J_\lambda| 0^+ \rangle = \left\{ -\epsilon_{\lambda\rho\sigma} h_{\rho\kappa} \left(\frac{q_\kappa}{2m_p}\right) \left(\frac{q_\sigma}{2m_p}\right) F_M(q^2) + ih_{\lambda\rho} \frac{q_\rho}{2m_{N_i}} F_A(q^2) \right.$$

$$\left. + iq_\lambda h_{\rho\sigma} q_\rho q_\sigma F_p(q^2)/(2m_{N_i} m_\pi^2) \right\} \tag{2}$$

Here $q$ is the momentum transfer 4-vector, $m_p$ proton mass, $m_{N_i}$ the mass of $O^{16}$, $\epsilon_{\lambda\rho\sigma}$ is a completely antisymmetric symbol with $\epsilon_{123} = +1$, and $h$ is the polarization tensor for spin 2. Invoking CVC, one can relate the form factor $F_M$ to the transverse magnetic form factor obtainable (if isospin is a good quantum number) from the inelastic electron scattering

$$e^- + O^{16}(0^+) \rightarrow e^- + O^{16*}(2^-, T = 1) \tag{3}$$

From PCAC follows that

$$F_p(q^2) = -\frac{m_\pi^2}{m_\pi^2 + q^2} F_A(q^2) \tag{4}$$

Next we can show that $F_A(0)$ can be obtained from the $\beta$-decay

$$N^{16}(2^-, T = 1) \rightarrow O^{16}(0^+, T = 0) + e^- + \bar{\nu}_e \tag{5}$$

since the terms multiplied by $F_M$ and $F_p$ vanish in the limit $q \rightarrow 0$ as is the case for $\beta$-decay processes. In order to deduce $F_A(q^2)$ for $q^2 \neq 0$ from $F_A(0)$, we assume the following $q^2$-dependence of the form factor

$$\frac{F_M(q^2)}{F_M(0)} = \frac{F_A(q^2)}{F_A(0)} \tag{6}$$

This can be verified in the impulse approximation. Equation (6) combined with the CVC and the PCAC equation (4) is sufficient to give the capture rate for (1). Similar analysis can be made for the process

$$\mu^- + C^{12}(0^+) \rightarrow \nu_\mu + B^{12}(1^+, T = 1) \tag{7}$$

The results obtained by Kim and Primakoff are quite encouraging, i.e., in unit of $10^3 s^{-1}$:

$$\Lambda_\mu(C^{12} \rightarrow B^{12}) = 6.6 \pm 1.0 \left(6.75 \, {}^{+0.35}_{-0.75}\right) \tag{8}$$

$$\Lambda_\mu(O^{16} \to N^{16}) = 5.8 \pm 2.8(6.3 \pm 0.6, 7.87 \pm 0.76\ddagger) \qquad (9)$$

where the numbers in parentheses are experimental. It should be emphasized that in terms of the impulse approximation, the capture rates are very much dependent on the nuclear structure effect.

If we suppose that the good agreement with experiment is an indication that our hypotheses, equations (4) and (5), are valid, we illustrate below a hitherto unexplored idea on the photopion production from $O^{16}$:

$$\gamma + O^{16}(0^+) \to \pi^+ + N^{16}(0^-, 1^-, 2^-, 3^-; T = 1) \qquad (10)$$

in which all the final states are bound. The integrated cross section to all bound states of $N^{16}$ has been studied experimentally. It was found in the theoretical study of Devanathan et al.[6] that the cross sections obtained with the well-accepted dispersion theoretic amplitude of Chew et al. combined with the impulse approximation are systematically too large by more than a factor of 2 compared to experiments even when reasonably accurate nuclear wave functions are used. This indicates that some mechanism like partial absorption of produced mesons must be taking place, probably inside the nucleus. Since the final states are exactly the same as those from the capture of muons, the partial transitions in the process (10) must be experimentally measurable as they are in muon capture. An interesting question is how much deviation each partial transition will show between theory and experiment. That will give us more solid starting point for a theoretical analysis. For the particular transition $\gamma + O^{16}(0^+) \to \pi^+ + N^{16}(2^-, T = 1)$, we have at our disposition the form factor $F_A(q^2)$ [in principle for all $q^2$ if we assume equation (6) holds]. Let us invoke the PCAC applicable in the presence of the electromagnetic field $\mathscr{A}_\mu$,

$$\partial_\mu J_\mu^A(x) + ie\mathscr{A}_\mu J_\mu^A(x) = ic\Phi_{\pi^+}(x) \qquad (11)$$

where $J_\mu^A$ is the axial vector current, $\Phi_{\pi^+}$ a pion field operator which creates $\pi^+$ and $c$ is a constant which can be related to the weak axial form factor. By taking the matrix element between $\langle N^{16}(2^-)|$ and $|\gamma, O^{16}(0^+)in\rangle$, we have

$$c\frac{1}{m_\pi^2 + q^2}\langle 2^-|J_{\pi^+}(0)|\gamma, 0^+ in\rangle = -q_\mu\langle 2^-|J_\mu^A(0)|\gamma, 0^+ in\rangle$$
$$+ e\langle 2^-|\mathscr{A}_\mu(0)J_\mu^A(0)|\gamma, 0^+ in\rangle \qquad (12)$$

---

‡The most recently measured value by J. P. Deutsch et al., Phys. Letters, to be published. I am grateful to Professor Deutsch for the communication prior to its publication.

where we have used the field equation $(-\Box + m_\pi^2)\,\Phi_{\pi^+}(x) = J_{\pi^+}(x)$, and $J_{\pi^+}$ is the source of $\pi^+$ field. The term on the left-hand side is exactly the matrix element required for the photoproduction process, and thus if we can suitably take care of the photon on the right-hand side, then we can express the photoproduction matrix element in terms of the *nuclear* weak axial-vector form factor $F_A(q^2)$. This sort of analysis has been done by Gaffney[7] for the charged pion production at thereshold from proton and is found to give good results. A similar calculation for $O^{16}$ using the idea of Kim and Primakoff is in progress. The success of this approach would be a support to the validity of the PCAC assumption and the "elementarity" of $O^{16}$ and $N^{16}$ ground states. It is anticipated that an experiment on the partial transitions of (10), the impulse approximation and the EPM results will shed some light on the discrepancy.

## 4. NUCLEAR WAVE FUNCTION

Now we ask what sorts of nuclear structure information can the $\mu$-capture process provide? Let us analyze in terms of nuclear wave functions the process‡

$$\mu^- + O^{16}(0^+) \to \nu_\mu + N^{16}(0^-, 1^-, 2^-, 3^-) \tag{13}$$

Except for the $3^-$ state, the partial transition rates have been measured experimentally. The transitions to $0^-$ and $2^-$ exhibit the dependence on the pseudoscalar coupling constant, while those to $1^-$ and $3^-$ do not. For our purpose, we assume the PCAC in the $0^-$ and $2^-$ transitions. We neglect the exchange current contribution. Then the capture rate for the transition $i \to f$ is

$$\Lambda_\mu(i \to f) = \text{const.} \sum_{M_f} \left\{ G_V^2 \Big| \int 1 \Big|^2 + G_A^2 \Big| \int \sigma \Big|^2 \right.$$
$$\left. + (G_p^2 - 2G_p G_A) \Big| \int \hat{\nu}\cdot\sigma \Big|^2 + \text{R. C.} \right\} \tag{14}$$

Here R.C. stands for relativistic corrections among which the most important one is the pseudoscalar-coupling dependent term

$$- \frac{2g_A(G_A - G_p)}{m_p} \left( \hat{\nu}\cdot\int\sigma \right)^* \left( \int \mathbf{p}\cdot\sigma \right) \tag{15}$$

‡The subsequent discussion is based on my lectures deliverd at the 1967 Summer Institute of Nuclear and Particle Physics, McGill University, Montreal, Canada (to be published); also M. Rho, *Phys. Rev. Letters* **18**: 671 (1967); *idid.*, **19**: 248 (1967); *Phys. Rev.* **161**: 955 (1967).

The matrix elements are defined by

$$\int y = \langle f | \sum_j \tau_j^{(-)} e^{-i\nu \cdot r_j} \varphi(r_j)\, y_j | i \rangle$$

$\varphi_\lambda$ is the $\mu$ wave functions, $\nu$ the neutrino energy, and $g$'s and $G$'s are coupling constants. The symbols are exactly the same as used by Foldy and Walecka.[8]

From the point of view of nuclear structure, there are three different kinds of transition operators, $\tau^{(-)}, \tau^{(-)}\sigma$, and $\tau^{(-)}\mathbf{p}$. If one is to restrict the consideration to the simplest configuration space, for example, one particle-one hole space for the final states, then it is necessary to take into account in the transition operators their renormalization in the presence of more complex configurations. Such a renormalization can be most easily considered in the framework of Migdal's approach to nuclear structure.[9] In this approach, the renormalization effect appears as something akin to the effective charge and therefore the modified operator $\mathcal{T}^\omega(t)$ for an operator $t$ may be written

$$\mathcal{T}^\omega(t) = e(t)t \tag{16}$$

where $e(t)$ is the generalized effective charge. Migdal has shown that the effective charge for $\tau^{(-)}$ is unity and those for $\tau^{(-)}\sigma$ and $\tau^{(-)}\mathbf{p}$ differ from unity. In fact his analysis of various experiments indicates that $e(\tau^{(-)}\sigma) < 0.90$ and $e(\tau^{(-)}\mathbf{p}) < 1$. Moreover, the Migdal's approach can be used to show that $\tau^{(-)}$ and $\tau^{(-)}\sigma$ can mainly excite the modes in the nuclei which are mediated primarily via the spin independent $(f_0 \tau_1 \cdot \tau_2)$ and spin dependent $(g_0' \tau_1 \cdot \tau_2 \sigma_1 \cdot \sigma_2)$ interactions respectively. In the same way, $\tau^{(-)}\mathbf{p}$ term would feel the velocity dependent term in Migdal's effective interaction. This last is assumed to be small.

Reality must be a lot more complicated than that, but one can gain considerable insight by analyzing the processes (13) in terms of the quantities mentioned above.

a) $0^+ \to 0^-$; Here the first term of equation (14) vanishes, and furthermore $|\int \sigma|^2 = |\int \hat{\nu} \cdot \sigma|^2$. For the Goldberger–Treiman value of $F_p, G_p$ is very near $G_A$; hence, the nonrelativistic contribution is largely suppressed. Thus the dominant contribution is equation (15). From the arguments given above, we see that the transition is independent of $f_0'$ and depends insensitively on $g_0'$. It will, however, be quadratic in $e(\tau^{(-)}\sigma)$ and feel more sensitively the velocity-dependent component. We conclude that as long as the velocity-dependent component is small, the capture rate will not be too sensitive to nuclear parameters:

b) $0^+ \rightarrow 1^-$; Here all the terms independent of $G_p$ contribute and, hence, the transition will feel both $f_0'$ and $g_0'$ components. But it turns out to be insensitive to those paramenters.

c) $0^+ \rightarrow 2^-$; Again $|\int 1|^2 = 0$. But $|\int \sigma|^2 \gg |\int \hat{\nu}\cdot\sigma|^2$ and hence the cancellation of the kind observed in the $0^-$ transition does not occur. Thus this transition is a candidate for determining $g_0'$ and checking the renormalization $e(\tau^{(-)}\sigma) \neq 1$.

The situation with theoretical calculations is as follows: a particle-hole model in an approximation scheme such as the RPA (i.e., Gillet wave function) can reasonably explain the first two processes, but gives too large capture rate for the last transition. It is clear that the deformation of the $O^{16}$ ground state cannot be the explanation. In our present model, the explanation is simple. That the two first processes are easily fit is clear from their insensitivity to $f_0'$ and $g_0'$. To find the cause of discrepancy in the last process, we should examine its dependence on $g_0'$. It is shown in unit of $10^3$ $s^{-1}$ in Table I.

**Table I**

Dependence of Capture Rates on $g_0'$

| $g_0'$ | $\Lambda(2^-)$ | $\Lambda(0^-)$ |
|---|---|---|
| 0 | 22.34 | 1.82 |
| 0.3 | 10.60 | 1.43 |
| 0.5 | 6.36 | 1.26 |
| 0.7 | 3.98 | 1.13 |

Notice the large variation of the $2^-$ capture rate as a function of $g_0'$. The reason why the former calculations failed by a factor of 2 or more can be found in the weakness of their spin-spin interaction $g_0'$. The conventional forces do not give such large $g_0'$. This large $g_0'$ is consistent with the electron scattering form factors and the inverse $\beta$-decay matrix elements. Roughly speaking, the force that Gillet uses has the ratio $g'_0/f'_0 = 2/3$, while the ratio which gives all 3 transitions correctly is $g'_0/f'_0 \approx 10/7$ (the magnitude of $f'_0$ is obtained from Migdal's analysis[9]).

This suggests that by modifying the current effective force so that $g'_0 \approx 0.5$ one can hope to explain all experiments without destroying other features. The fundamental nuclear structure problem is then: If indeed the $g'_0$ amplitude is the cause of the discrepancy, can one simulate such a strong spin-dependence through a Kuo–Brown type calculation[10] starting from the nucleon–nucleon potential? I do not have the answer, but so far the Kuo–Brown calculation has not obtained it.

## 5. SUPERMULTIPLET SYMMETRY

One more consequence of the strong spin-dependence which seem to be necessary above and of the particular renormalization of the $\sigma$-operator is that the $SU(4)$ relation

$$M_V^2 = M_A^2 = M_p^2 \tag{17}$$

where

$$M_{V,A,p}^2 = \sum_f \left(\frac{v_f}{m_\mu}\right)^2 \int \frac{d\hat{v}}{4\pi} |\langle f| \sum_j \tau_j^{(-)} e^{-i\nu \cdot r_j} O_{V,A,p}|i\rangle|^2$$

$$O_V = 1, \quad O_A = \frac{1}{\sqrt{3}}\sigma, \quad O_p = \hat{v} \cdot \sigma$$

cannot hold anymore. The relationship (17) has always been assumed in most of muon capture calculations,[10] and actually holds within 12% in all *other* models. The deviation from the above equality has been estimated for $O^{16}$ and can be calculated easily from Bunatyan's paper.[9] On the average,

$$0.70 \lesssim \frac{M_A^2}{M_V^2} \lesssim 0.75 \tag{18}$$

It should be noted that this is a large deviation from 1.13 which a conventional particle-hole model would predict. It can also be shown that the deviation indicated in equation (18) leads to total capture rates in better agreement with experiments.[9] If equation (18) is the true effect, then the $SU(4)$ supermultiplet symmetry may not be such a good one.

## REFERENCES

1. C. S. Wu, *Rev. Mod. Phys.*, **36**: 618, (1964).
2. V. Devanathan and M. E. Rose, University of Virginia, Document ORO-2915-81.
3. M. Conversi, R. Diebold, and L. Di Lella, *Phys. Rev.* **136**: B1077, (1964).
4. Private Communication from C. W. Kim; S. L. Adler, and Y. Dothan, *Phys. Rev.* **151**: 1267 (1966).
5. C. W. Kim, *Phys. Rev.* **151**: 1261 (1966) where prior references can be found; also H. Primakoff, "Tokyo Lectures on High-Energy Physics," Ed. A. Fujii, Syokabo and Benjamin, 1967.
6. V. Devanathan, M. Rho, K. Srinivasa Rao, and S. C. K. Nair, *Nucl. Phys.* **B2**: 329 (1967).
7. G. W. Gaffney, *Phys. Rev.* **161**: 1599 (1967).
8. L. Foldy and J. D. Walecka, *Nuovo Cim.* **34**: 1026 (1964).
9. A. B. Migdal, "Proceedings of the International School of Physics Enrico Fermi," Academic Press, 1966.
10. T. T. S. Kuo and G. E. Brown, *Nucl. Phys.* **A92**: 481 (1967).

# Functional Differential Equations

## S. G. DEO

*MARATHWADA UNIVERSITY*
*Aurangabad, India*

---

## 1. INTRODUCTION

Lyapunov's second method gives sufficient conditions for stability and asymptotic stability. This method has been extended in several directions.[7,8] One of the interesting extension of this method depends basically on the fact that a function satisfying the inequality

$$m'(t) \leq w(t, m(t)) \qquad m(t_0) = r_0 \tag{1}$$

is majorized by the maximal solution of the equation

$$r' = w(t, r) \qquad r(t_0) = r_0$$

This comparision principle has been successfully employed to study a variety of problems in ordinary differential equations.

Below we use this comparision principle and extension of Lyapunov's method to study some problems in functional differential equations. The study of these types of equations has been done in considerable detail and a good deal of literature on this problem is now available.[3-5,7-10]

We consider the functional differential equation as formulated by Hale.[4] It is known[2,4] that when we use Lyapunov functionals we require the form of the solution of the functional differential equation to define the derivative of the same. As is known,[2,7,8,10] one can study the functional differential equation by the use of

Lyapunov function, in which case, the knowledge of solution is not needed and, therefore, a basic problem arises in the selection of minimal set of functions, along which the Lyapunov function must satisfy the assumptions. In the latter case, it is convenient to consider the solution as an element of the function space at $t = t_0$ and thereafter as an element of $R^n$ for $t > t_0$. This leads to the notion of shift-invariancy. We define a shift-invariant set with respect to a given set and obtain sufficient conditions for the stability and boundedness of the same by using a Lyapunov function. As is expected, the results cover the existing notions, as special cases, and are therefore of interest.

It is natural to expect that a corresponding analogous result to (1) for functional differential equations would be equally useful. If we use this generalized comparision principle, then, the above requirement of the minimal set of functions can be overcome. We prove below the comparision theorem for a more general type of functional differential equation.

*Notation:* Let $I$ denote the interval $0 \leq t < \infty$ and $R^n$, $n$-dimensioal Euclidean space. For any vector $x \in R^n$, let $\|x\|^2 = \sum_{i=1}^{n} x_i^2$. Let $C$ denote the space of continuous functions from $[-\tau, 0]$ into $R^n (\tau > 0)$. For any $\phi \in C$, define

$$|\phi| = \max_{-\tau \leq s \leq 0} \|\phi(s)\|$$

Given a continuous function $x(u)$ from $[-\tau, \infty)$ into $R^n$ and any fixed $t, 0 \leq t < \infty$, define $x_t \in C$ by $x_t = x_t(s) = x(t + s), (-\tau \leq s \leq 0)$; that is, $x_t \in C$ is that "segment" of the function $x(u)$ defined by letting $u$ range in the interval $t - \tau \leq u \leq t$. With this notation, we may write the functional differential system in the vector form

$$x'(t) = f(t, x_t) \tag{2}$$

where $f(t, \phi)$ is defined and continuous an $I \times C$ and $f$ maps $I \times C$ into $R^n$. Here and in what follows (') denotes the right-hand derivative.

*Definition:* Let $\phi \in C$ be any given function and $t_0 \geq 0$. Any continuous functions $x(t_0, \phi)(t)$ from $[t_0 - \tau, \infty)$ into $R^n$ is said to be a solution of (2), if
1. $x_t(t_0, \phi) \in C$ for all $t \geq t_0$
2. $x_{t_0}(t_0, \phi) = \phi$
3. $x(t_0, \phi)(t)$ exists for all $t \geq t_0$ and satisfies (2)

## 2. SHIFT INVARIANT SETS

Let $A$ be any subset of $C$ and $B$ any subset of $R^n$ and $x(t_0, \phi)$ be any solution of (2).

*Definition:* The set $B$ is said to be shift invariant with respect to the set $A$ if $\phi \in A$ (at $t = t_0$) implies that $x(t_0, \phi)(t) \subset B$ for all $t > t_0$.

Let $g$ be a vector of $m$-dimensions and the function $g(x)$ be defined and continuous on $R^n$. Define

$$\|g(x)\|^2 = \sum_{i=1}^{m} |g_i(x)|^2$$

and

$$|g(\phi)| = \max_{-\tau \leq s \leq 0} \|g(\phi(s))\|$$

Let the sets $A$ and $B$ be defined as follows:

$$A = \{\phi \in C; |g(\phi)| = 0\}$$
$$B = \{x \in R^n, \|g(x)\| \leq \alpha\}$$

where $\alpha$ is some positive real number. Let the set $B$ be shift invariant with respect to the set $A$. We need the following sets:

$$\bar{N}(A, \eta) = \{\phi \in C; |g(\phi)| \leq \eta\}$$
$$S(B, \eta) = \{x \in R^n; \|g(x)\| \leq \alpha + \eta\} \qquad (\eta \geq 0)$$

We list below the following definitions which unify our results on stability and boundedness of the shift invariant set $B$ with respect to $A$ and the system (2):

(i) Given $\epsilon > 0$ and $t_0 \geq 0$, there exists a positive function $\delta(t_0, \epsilon, \alpha)$ that is continuous in $t_0$, for each $\epsilon$, such that

$$x(t_0, \phi)(t) \subset S(B, \epsilon) \qquad (t > t_0)$$

whenever

$$\phi \in \bar{N}(A, \delta)$$

(ii) The $\delta$ in (i) is independent of $t_0$.

(iii) Given $\epsilon > 0$, $\gamma > 0$ and $t_0 \geq 0$, there exists a positive number $T = T(t_0, \alpha, \gamma, \epsilon)$ such that

$$x(t_0, \phi)(t) \subset S(B, \epsilon) \qquad (t \geq t_0 + \tau)$$

whenever

$$\phi \in \bar{N}(A, \gamma)$$

(iv) The $T$ in (iii) is independent of $t_0$.

(v) Definitions (i) and (iii) hold simulteneously.

(vi) Definitions (ii) and (iv) hold simulteneously.

(vii) Given $\gamma > \alpha$ and $t_0 \geq 0$, there exists a postive function $\eta = \eta(t_0, \gamma, \alpha)$ that is continues in $t_0$ for each $\gamma$ such that

$$x(t_0, \phi)(t) \subset S(B, \eta) \qquad t > t_0$$

whenever

$$\phi \in \bar{N}(A, \gamma)$$

(viii) The $\eta$ in (vii) is indepedent of $t_0$.

(ix) For each $\gamma > \alpha$ and $t_0 \geq 0$ there exist positive number $\mu$ and $T = T(t_0, \gamma, \alpha)$ such that

$$x(t_0, \phi)(t) \subset S(B, \mu) \qquad (t \geq t_0 + T)$$

whenever

$$\phi \in \bar{N}(A, \gamma)$$

(x) The $T$ in (ix) is independent of $t_0$.

(xi) Definitions (vii) and (ix) hold simultaneously.

(xii) Definitions (viii) and (x) hold simultaneously.

*Remarks:* If $\alpha = 0$ and $A = \{\theta\}$ where $\theta$ is the null element of $C$ and $B = \{0\}$ then we obtain the stability and boundness of the trivial solution of (2).

Let $\tilde{C}$ denote the space of continuous functions from $[t_0 - \tau, \infty)$ into $R^n$. Let $V(t, x) \geq 0$ be defined and continuous on $[t_0 - \tau, \infty) \times R^n$ and suppose that it satisfies Lipschitz's condition locally in $x$. Define

$$V^*(t, x(t)) = \lim_{h \to 0^+} \sup \frac{1}{h} [V(t + h, x(t) + hf(t, x_t)) - V(t, x(t))]$$

$$\tag{3}$$

for each $t \in [t_0, \infty)$ and $x \in \tilde{C}$.

$$\tilde{C}(A) = \{x \in \tilde{C} : \sup_{t-\tau \leq s \leq t} V(s, x(s)) A(s) = V(t, x(t)) A(t)\} \tag{4}$$

where $A$ is positive scalar function defined and continuous on $[t_0 - \tau, \infty)$ and $A'(t)$ exists for each $t \in [t_0, \infty)$.

Let $W(t, u) \geq 0$ be defined and continuous for $t \in [t_0, \infty)$ and $u \geq 0$. We are interested in the following differential equations:

$$u' = W(t, u) \qquad u(t_0; t_0, u_0) = u_0 \tag{5}$$

$$r' = \{-A'(t) r + W(t, A(t)r)\} [A(t)]^{-1} \qquad r(t; t_0, r_0) = r_0 \tag{6}$$

If $u(t; t_0, u_0)$ and $r(t; t_0, r_0)$ are the maximal solutions of (5) and (6), respectively, then by using the method of variation of parameters one finds that

$$u(t; t_0, u_0) = r(t; t_0, r_0) A(t) \qquad (7)$$

whenever

$$u(t_0) = r(t_0) A(t_0)$$

We next state the following lemma which is a modification of a similar lemma proved elsewhere.[6] The proof of the lemma can be formulated with necessary changes.

*Lemma:* Let the function $W(t, u) \geq 0$ be defined as above and $r(t; t_0, r_0)$ be the maximal solution of (6) existing to the right of $t_0$. For each $t \in [t_0, \infty]$ and $x(t) \in \tilde{C}(A)$, let

$$A(t) V^*(t, x(t)) + A'(t) V(t, x(t)) \leq W(t, A(t) V(t, x(t))) \qquad (8)$$

If $x(t_0, \phi)$ is any solution of (2) with the initial function $\phi$ existing for all $t \geq t_0$ such that

$$V(t_0, \phi(s)) \leq r_0 \qquad (t_0 - \tau \leq s \leq t_0) \qquad (9)$$

then

$$V(t; x(t_0, \phi)(t)) \leq r(t; t_0, r_0) \qquad (t \geq t_0)$$

## 3. STABILITY AND BOUNDEDNESS THEOREMS

Let $r(t; t_0, r_0)$ be a solution of (6). If $r_0 = 0$ assume that

$$r(t; t_0, r_0) \leq \beta(t_0) = \beta \qquad (t \geq t_0) \qquad (10)$$

Now we need the definitions (i*) to (xii*) which the differential equation (6) satisfy. They correspond to the definitions (i) to (xii) given earlier. We state below (i*) and (ii*) and omit the others which can be formulated similarly.

(i*) Given $\epsilon > 0$ and $t_0 \geq 0$ there exists a positive function $\delta = \delta(t_0, \epsilon)$ that is continuous in $t_0$ for each $\epsilon$, such that

$$r(t; t_0, r_0) < \beta + \epsilon \qquad (t \geq t_0)$$

whenever

$$r_0 \leq \delta$$

where $\beta$ is as defined in (10).

(ii*) The $\beta$ in (10) and $\delta$ in (i*) are both independent of $t_0$.

We further assume that

$$V(t, \phi(s)) = 0 \qquad \text{if and only if } \phi \in A \tag{11}$$

The function $b(u)$ is defined, continuous nondecreasing in $u$, for $u \geq 0, b(u) > 0$ for $u > 0$ and is such that

$$b(\|g(x)\|) \leq V(t, x) \tag{12}$$

for each $(t, x) \in [t_0, \infty) \times R^n$. Let $b(u) \to \infty$ as $u \to \infty$.

$$V(t, x) \to 0 \text{ as } \|g(x)\| \to 0 \tag{13}$$

uniformly in $t_0$. We now state the following theorems on stability and boundedness of the shift invariant set.

*Theorem 1:* Let the assumptions of the lemma hold together with (11), (12), and (13). Then

(I) Condition (i*) $\Rightarrow$ condition (i)

(II) Condition (iii*) $\Rightarrow$ condition (iii)

(III) Condition (v*) $\Rightarrow$ condition (v)

*Proof:* Let $x(t_0, \phi)$ be any solution of (1) such that

$$V(t_0, \phi(s)) \leq r_0 \tag{14}$$

Since the assumptions of the lemma hold, we have in view of (3.5)

$$V(t; x(t_0, \phi)(t)) \leq r(t; t_0, r_0) \qquad (t > t_0) \tag{15}$$

Choose $r_0 = 0$, as $V(t, x) \geq 0$ we conclude from (11) and (14) that $\phi \in A$. Since (12) holds, (15) together with (10) implies that

$$\|g(x(t_0, \phi)(t))\| \leq b^{-1}(\beta) \equiv \alpha \qquad \text{for } t > t_0 \tag{16}$$

It then follows that whenever $\phi \in A$, $\|g(x(t_0, \phi)(t))\| \leq \alpha$, which means that $x(t_0, \phi)(t) \subset B$ for $t > t_0$. Thus $B$ is the shift invariant set with respect to the set $A$.

Now let $\epsilon > 0$ be given. Since (i*) holds, given $b(\alpha + \epsilon) > 0$ and $t_0 > 0$, where $\alpha$ is the same number as defined in (16), there exists a positive function $\delta = \delta(t_0, \epsilon)$ that is continuous in $t_0$ for each $\epsilon$, such that

$$r(t; t_0, r_0) < b(\alpha + \epsilon) \qquad t \geq t_0 \tag{17}$$

whenever

$$r_0 \leq \delta \tag{18}$$

Since (13), (14), and (15) hold there exists a positive number $\delta_1$ such that

$$\sup_{|\sigma(\phi)| \leq \delta_1} V(t_0, \phi(s)) \leq \delta \tag{19}$$

Since $|g(\phi)| \leq \delta_1$ implies that $\phi \in \bar{N}(A, \delta_1)$, it follows that whenever $\phi \in \bar{N}(A, \delta_1)$ every solution $x(t_0, \phi)$ satisfies (15).

Suppose, if possible, that a solution $x(t_0, \phi)$ of (1) is such that $x(t_0, \phi)(t_1) \not\subset S(B, \epsilon)$ for some $t = t_1 > t_0$ whenever $\phi \in \bar{N}(A, \delta_1)$. Then using the relations (12), (15), and (17) we get the contradiction

$$b(\alpha + \epsilon) \leq V(t_1, x(t_0, \phi)(t_1)) \leq r(t_1; t_0, r_0) < b(\alpha + \epsilon)$$

which proves (i).

Now to prove that (iii*) implies (iii), let $\epsilon > 0$, $\gamma > 0$ and $t_0 \geq 0$ be given. Suppose that $|g(\phi(s))| \leq \gamma$. Then because of (13), there exists an $\alpha_1 = \alpha_1(\gamma)$ such that

$$\sup_{|\sigma(\phi(s))| \leq \gamma} V(t_0, \phi) \leq \alpha_1 \qquad (20)$$

We choose $r_0 \leq \alpha_1$. Then because of (14) whenever $\phi \in \bar{N}(A, \gamma)$, every solution $x(t_0, \phi)(t)$ of (1) satisfies (15).

Since (iii*) holds, given $b(\alpha + \epsilon) > 0$, $\alpha_1 > 0$ and $t_0 \geq 0$ there exists a positive number $T = T(t_0, \alpha_1, \epsilon)$ such that

$$r(t; t_0, r_0) < b(\alpha + \epsilon) \qquad (t \geq t_0 + T) \qquad (21)$$

whenever

$$r_0 \leq \alpha_1$$

Let $\{t_n\}$ be a sequence such that $t_n \to \infty$ as $n \to \infty$ and $t_n \geq t_0 + T$ for each $n$. Suppose that a solution $x(t_0, \phi)$ of (1) such that $\phi \in \bar{N}(A, \gamma)$ has the property that $x(t_0, \phi)(t_n) \not\subset S(B, \epsilon)$. Since $b(u)$ is monotonic and (12), (15), and (21) hold, we arrive at the contradiction

$$b(\alpha + \epsilon) \leq V(t_n, x_{t_n}(t_0, \phi)) \leq r(t_n, t_0, r_0) < b(\alpha + \epsilon)$$

which proves (iii).

If the conditions (i*) and (iii*) hold simultaneously then by combining the above proofs it follows that the condition (v*) implies (v). This completes the proof of theorem 1.

*Theorem 2:* Let the hypothesis of theorem 1 hold. Then

     (I) Condition (iia) $\Rightarrow$ condition (ii)

     (II) Condition (iva) $\Rightarrow$ condition (iv)

     (III) Condition (via) $\Rightarrow$ condition (vi)

*Proof:* Since (ii*) holds, $\beta$ of (10) and $\delta$ of (18) are independent of $t_0$ and therefore the same is true of $\alpha_1$ and $\delta_1$ of (16) and (19), respectively. Now the proof of the conclusion (ii*) implies (ii) is essentially the same as that of theorem 1. The other conclusions follow similarly.

*Theorem 3:* Let the hypothesis of theorem 1 hold. Then

  (I) Condition (vii*) $\Rightarrow$ condition (vii)
  (II) Condition (ix*) $\Rightarrow$ condition (ix)
  (III) Condition (xi*) $\Rightarrow$ condition (xi)

*Proof:* Let $\gamma > \alpha$ and $t_0 \geq 0$ be given. We proceed as in Theorm 1 and conclude that whenever $\phi \in \bar{N}(A, \gamma)$ every solution $x(t_0, \phi)$ of (1) satisfies (15).

Since (vii*) holds, given $\alpha_1 > 0$ and $t_0 \geq 0$ there exists $\eta = \eta(t_0, \alpha_1)$, such that

$$r(t; t_0, r_0) < \beta + \eta \qquad (t \geq t_0) \tag{22}$$

whenever

$$r_0 \leq \alpha_1$$

where $\alpha_1$ is defined in (20). Also because $b(u) \to \infty$ as $u \to \infty$, there exists a $L = L(t_0, \alpha_1)$, Such that

$$\beta + \eta \leq b(\alpha + L) \tag{23}$$

Suppose that there is a $t_1 > t_0$ such that whenever $\phi \in \bar{N}(A, \gamma)$ and $x(t_0, \phi)(t_1) \not\subset S(B, L)$. Since (12), (15), (22), and (23) hold, we get

$$b(\alpha + L) \leq V(t_1, x(t_0, \phi)(t_1)) \leq r(t_1; t_0, r_0) < \beta + \eta \leq b(\beta + L)$$

This contradiction proves (vii).

The proofs of the remaining statements follow closely the proofs of theorem 1 and the one given above. We leave the details.

*Theorem 4:* Let the hypothesis of theorem 1 hold. Then

  (I) Condition (viii*) $\Rightarrow$ condition (viii)
  (II) Condition (x*) $\Rightarrow$ condition (x)
  (III) Condition (xii*) $\Rightarrow$ condition (xii)

*Proof:* The proofs of these conclusions can be constructed from the proofs of the above theorems. The details are omitted.

*Example:* We consider the functional differential equation.

$$x'(t) = \tfrac{1}{2}(\cos t - e^{\sin t} - e^{1/2\sin t - t}) x(t)$$

$$+ e^{1/2\sin t - t}\left[x^2(t) + e^{\sin t}\int_{-\tau}^{0} x^2(t + s)\,ds\right]^{1/4} \qquad (\tau > 0)$$

Taking $V(t, x) = x^2$ and $g(x) = x$, we get $b(u) = u^2$, which satisfies our requirements. Thus the set $A = \{\theta\}$ where $\theta$ is the null element of $C$. Clearly, condition (11) holds. From (4), when $A(t) \equiv 1$, we have

$$\tilde{C}(A) = [x \in \tilde{C}: \sup_{-\tau \leq s \leq 0} x^2(t + s) = x^2(t)]$$

Now, we observe that

$$V^*(t, x(t)) \leq (\cos t - e^{\sin t} - e^{1/2\sin t - t}) V$$
$$+ e^{1/2\sin t - t} \left[ x^2(t) + \{x^2(t) + e^{\sin t} \int_{-\tau}^{0} x^2(t + s)\, ds\}^{1/2} \right]$$

when $x(t) \in \tilde{C}(A)$,

$$V^*(t, x(t)) \leq \cos t V + e^{1/2\sin t - t} V^{1/2} (1 + e\tau)^{1/2}$$

The maximal solution of (6) is then given by

$$r(t; t_0, r_0) = [r_0 e^{1/2(\sin t - \sin t_0)} + e^{\sin t} \{e^{-t_0} - e^{-t}\} (1 + e\tau)^{1/2}]^2$$

Putting $r_0 = 0$, we get in view of (10) that $\beta = e^{1 - 2t_0}(1 + e\tau)$. From the relation (16), it is easily seen that $\alpha = e^{1/2 - t_0}(1 + e\tau)^{1/2}$. Hence $B = [x \in R^1; \|x\| \leq e^{1/2 - t_0}(1 + e\tau)^{1/2}]$ is the shift-invariant set with respect to the set $A = \{\theta\}$. It is clear that the condition (i*) holds and therefore by theorem 1 we conclude that (i) holds.

## 4. FINITE SYSTEMS OF FUNCTIONAL DIFFERENTIAL INEQUALITIES

Let $C_H$ denote the set of functions $\phi \in C$ such that $|\phi| \leq H$ ($H > 0$). Below we consider a more general functional differential system

$$x'(t) = f(t, x(t), x_t) \tag{24}$$

where $f(t, x, \phi)$ is a functional which is defined and continuous on $I \times R^n \times C_H \to R^n$. The following theorem asserts the existence of the solutions of (24).

*Theorem 5:* Let $f(t, x, \phi)$ be defined and continuous on $I \times R^n \times C_H$. Then, for a given initial function $\phi \in C_H$ at $t = t_0$, there exists an $\alpha > 0$, such that there is a solution $x(t_0, \phi)(t)$ of (24) on $[t_0, t_0 + \alpha)$.

One can easily prove this theorem using the Shauder's fixed point theorem as in Ref. 11 and, hence, the proof is deleted. We state below a theorem which gives sufficient conditions for any solution of (24) to exist for all $t \geq t_0$. The proof is essentially similar to that in Ref. 11.

*Theorem 6:* Let the function $w(t, u, v)$ be defined and continuous on $I \times R^+ \times R^+$ where $R^+$ denotes the non-negative real line. Let $w$ be nondecreasing in $u$ and $v$ for each $t \in I$. Assume that the maximal solution $r(t)$ of the ordinary differential equation.

$$r' = w(t, r, r) \qquad r(t_0) = r_0 \geq 0 \tag{25}$$

exists for all $t \geq t_0$. Suppose further that

$$\|f(t, x, \psi)\| \leq w(t, \|x\|, |\psi|)$$

Then the largest interval of the existence of any solution $x(t_0, \phi)(t)$ of (24) is $[t_0, \infty)$.

*Definition:* We shall say that the function $f(t, x, \psi)$ possesses a quasi-monotone property if the following conditions hold: $f_p(t, x, \psi)$ is nondecreasing in $x_j$ $(j = 1, 2, \ldots, n)$ $j \neq p$ for each $(t, \psi)$ and also $f_p(t, x, \psi)$ is nondecreasing in $\psi_j$ $(j = 1, 2, \ldots, n)$ $j \neq p$ for each $(t, x)$.

We note in subsequent discussions that whenever vectorial inequalities will be used it is to be understood that the same inequalities hold between their corresponding components. Now we are in a position to state the following theorem which is an extension of a similar theorem in Ref. 11. This theorem plays an important role in the theory of functional differential inequalities since it is useful to establish many results.

*Theorem 7:* (a) Let the functional $f(t, x, \psi)$ be defined and continuous on $I \times R^n \times C_H$ and satisfy the quasimonotone property. (b) let $x(u)$ and $y(u)$ be continuous functions from $[t_0 - \tau, \infty) \to R^n$, such that $(t, x(t), x_t), (t, y(t), y_t) \in I \times R^n \times C_H$. Suppose further that $x_{t_0} < y_{t_0}$ and for $t > t_0$

$$D_- x(t) \leq f(t, x(t), x_t)$$
$$D_- y(t) > f(t, y(t), y_t)$$

then

$$x(t) < y(t) \qquad \text{for } t \geq t_0$$

The proof of the theorem follows by following a similar argument as in Ref. 11. We will omit the details.

*Definition:* Let $r(t_0, \phi)(t)$ be a solution of (24) existing on $[t_0, \infty]$. For any other solution $x(t_0, \phi)(t)$ of (24), existing on the same interval, if

$$x(t_0, \phi)(t) \leq r(t_0, \phi)(t) \qquad (t \geq t_0)$$

then $r(t_0, \phi)(t)$ is said to be the maximal solution of (24). A similar definition can be given for the minimal solution.

*Theorem 8:* Let the functional $f(t, x, \psi)$ be defined and continuous on $I \times R^n \times C_H$. Assume further that $f(t, x, \psi)$ satisfy the quasi-monotone property. Then, for a given function $\phi \in C_H$, there exists an $\alpha_1 > 0$ such that there is a maximal and a minimal solution of (24) on $[t_0, t_0 + \alpha_1)$.

*Proof:* Let $a$ and $b$ be positive real numbers such that $b \geq 2H$. Define the set $\Omega \subset I \times R^n \times C_H$

$$\Omega = \begin{cases} (t, x, \psi): & \text{(i)} \ t_0 \leq t \leq t_0 + a \\ & \text{(ii)} \ \|x - \phi(0)\| \leq b \quad x \in R^n \\ & \text{(iii)} \ |\phi - \phi(0)| \leq b \quad \psi \in \tilde{C}_H \end{cases}$$

where $\tilde{C}_H \subset C_H$ is the set of Lipschitz continuous functions with a suitable Lipschitz's constant. Since $f(t, x, \psi)$ is continuous on $\Omega$, which is a compact set in virtue of Ascoli–Arzala's theorem it is bounded there. Let $M$ be the bound of $f(t, x, \psi)$ on $\Omega$. Let $0 < \epsilon < b/2$. Consider

$$x'(t) = f(t, x(t), x_t) + \epsilon, \ x_{t_0}(t_0, \phi, \epsilon) = \phi + \epsilon \qquad (26)$$

It is obvious that $f_\epsilon(t, x, \psi) = f(t, x, \psi) + \epsilon$ is defined and continuous on $\Omega_\epsilon$, where

$$\Omega_\epsilon = \begin{cases} (t, x, \psi): & \text{(i)} \ t_0 \leq t \leq t_0 + a \\ & \text{(ii)} \ \|x - (\phi(0) + \epsilon)\| \leq b/2 \quad x \in R_n \\ & \text{(iii)} \ |\psi - (\phi(0) + \epsilon)| \leq b/2 \quad \psi \in \tilde{C}_H \end{cases}$$

Clearly $\Omega_\epsilon \subset \Omega$ and the bound of $f$ on $\Omega_\epsilon$ is $M + b/2$. Therefore by following the theorem on existence there exists a solution of (26) on $[t_0, t_0 + \alpha_1)$ where $\alpha_1 = \min(a, b/2M + b)$. For $0 < \epsilon_2 \leq \epsilon_1 \leq \epsilon$, we have

$$x_{t_0}(t_0, \phi, \epsilon_2) < x_{t_0}(t_0, \phi, \epsilon_1)$$

and

$$x'(t_0, \phi, \epsilon_2)(t) \leq f(t, x(t_0, \phi, \epsilon_2)(t), x_t(t_0, \phi, \epsilon_2)) + \epsilon_2$$
$$x'(t_0, \phi, \epsilon_1)(t) > f(t, x(t_0, \phi, \epsilon_1)(t), x_t(t_0, \phi, \epsilon_1)) + \epsilon_2$$

Thus theorem 7 can be applied to give from the above inequalitis

$$x(t_0, \phi, \epsilon_2)(t) < x(t_0, \phi, \epsilon_1)(t)$$

for $t \in [t_0, t_0 + \alpha_1)$

Since the family of functions $\{x(t_0, \phi, \epsilon)(t)\}$ is equi-continuons and uniformly bounded on $[t_0, t_0 + \alpha_1)$, as can be seen, it follows by Ascoli theorem, that there exists a decreasing sequence $\{\epsilon_n\}$, $\epsilon_n \to 0$ as $n \to \infty$, such that

$$\lim_{n \to \infty} x(t_0, \phi, \epsilon_n)(t) = r(t_0, \phi)(t)$$

uniformly on $[t_0, t_0 + \alpha_1)$. Clearly $r_{t_0}(t_0, \phi) = \phi$. The uniform continuity of $f(t, x, \psi)$ implies that $f(t, x(t_0, \psi, \epsilon_n)(t), x_t(t_0, \phi, \epsilon_n))$ tends uniformly to $f(t, r(t_0, \phi)(t), r_t(t_0, \phi))$ as $n \to \infty$ and then term by term integration is applicable to

$$x(t_0, \phi, \epsilon_n)(t) = \phi(0) + \epsilon_n + \int_{t_0}^{t} [f(s, x(t_0, \phi, \epsilon_n)(s), x_s(t_0, \phi, \epsilon_n)) + \epsilon_n]ds$$

which in turn implies that $r(t_0, \phi)(t)$ is a solution of (24) on $[t_0, t_0 + \alpha_1)$.

We shall now show that $r(t_0, \phi)(t)$ is a maximal solution of (24) on $[t_0, t_0 + \alpha_1)$. Let $x(t_0, \phi)(t)$ be any solution of (24) on $[t_0, t_0 + \alpha_1)$. Then, we have

$$x_{t_0}(t_0, \phi) < x_{t_0}(t_0, \phi, \epsilon) \qquad (27)$$

and

$$x'(t_0, \phi)(t) < f(t, x(t_0, \phi)(t), x_t(t_0, \phi)) + \epsilon \qquad (28)$$

$$x'(t_0, \phi, \epsilon)(t) \geq f(t, x(t_0, \phi, \epsilon)(t), x_t(t_0, \phi, \epsilon)) + \epsilon \qquad (29)$$

for $\epsilon \leq b/2$. By using theorem 7, it follows from (28) and (29), together with (27), that

$$x(t_0, \phi)(t) < x(t_0, \phi, \epsilon)(t) \qquad t \in [t_0, t_0 + \alpha_1) \qquad (30)$$

As a consequence of taking the limits as $\epsilon \to 0$ in (30), we have

$$x(t_0, \phi)(t) \leq \lim_{\epsilon \to 0} x(t_0, \phi, \epsilon)(t) = r(t_0, \phi)(t)$$

for $t \in [t_0, t_0 + \alpha_1)$. This proves the result.

We can prove the following comparision theorem in functional differential inequalities.

*Theorem 9:* Let the function $f(t, x, \psi)$ as defined in theorem 5 satisfy the quasi-monotone property. Suppose that the maximal solution $r(t_0, \phi)(t)$ of (24) exists on $[t_0, \infty)$. Let $m(u)$ be a continuous function $[t_0 - \tau, \infty) \to R^n$, $(t, m(t), m_t) \in I \times R^n \times C_H$ and $m_{t_0} \leq r_{t_0}$ $(t_0, \phi)$. Assume further that the inequality

$$D_- m(t) \leq f(t, m(t), m_t)$$

holds for $t \in [t_0, \infty) - S$, where $S$ is an almost countable subset of $[t_0, \infty)$, then

$$m(t) \leq r(t_0, \phi)(t)$$

hold for $t \geq t_0$. The proof follows from the proofs of theorems 7 and 8. We will omit the details.

# REFERENCES

1. L. P. Burten, and W. M. Whyburn, "Minimax solutions of ordinary differential equations." *Proc. Am. Math. Soc.* **3**: 794–803 (1952).
2. S. G. Deo, Ph. D. Thesis; submitted to Marathwada University, Aurangabad (Manarashtra) India.
3. J. Franklin, "On the existence of solution of system of functional differential equations." *Proc. Am. Math. Soc.* **5**: 363–369 (1954).
4. J. K. Hale, "Asymptotic behavior of the solutions of differential equations," Tech. Rep. 61-10 RIAS, Baltimore, 1961.
5. N. N. Krasovski, "Stability of Motion," Stanford University Press, Stanford, Calif.
6. V. Lakshmikantham, "On the boundedness of solutions of nonlinear differential equations," *Proc. Am. Math. Soc.* **8**: 1044–1048 (1957).
7. V. Lakshmikantham, "Lyapunov functions and a basic inequality in delay differential equations." *Arc. Rat. Mach. Anal.* **10**: 305–310 (1962).
8. V. Lakshmikantham, "Functional differential systems and extension of Lyapunov's method," *J. Math. Anal. Appl.* **8**: 392–405 (1964).
9. A. D. Myskis, "General theory of differential equations with retarded argruments (Russian)," *Uspehi. Mat. Nank.* **4**: 99–141 (1949).
10. B. S. Razumikhin, "On the stability of system with retardation," *Prik. Math. Mach.* **20**: 500–512 (1956).
11. G. R. Shendge and V. Lakshmikantham, "Functional differential inequalities; *An. da. Acad. Brazileira de Ciëncias* **39** (1): 31 (1967).
12. M. Oguztöreli Namik., "Time-lag Control Systems," Academic Press, New York, 1966.

# On Some Convex Functions and Related Inequalities

C. J. Eliezer

*UNIVERSITY OF MALAYA*
*Kuala Lumpur, Malaysia*

## 1. INTRODUCTION

In this paper I would like to present some work on inequalities, which Professor D. E. Daykin of the University of Malaya and I have recently completed.

Suppose we write Cauchy's inequality[1] in the form

$$F \equiv (\sum a^2)(\sum b^2) \geq (\sum ab)^2 \equiv G \tag{1}$$

where $a_1, \ldots, a_n$; $b_1, \ldots, b_n$ are non-negative real numbers. $F = G$ if the sets $(a)$ and $(b)$ are proportional.

An interesting problem is to consider whether one can find any expression of value intermediate between those of the two sides of the above inequality.

E. A. Milne[2] appears to be the first person to find one such expression. He showed that

$$\{\sum(a^2 + b^2)\} \sum \frac{a^2 b^2}{a^2 + b^2} \tag{2}$$

lies between $F$ and $G$.

Recently, Callebant[3] has shown that the expression

$$R = (\sum a^{1-x} b^{1+x})(\sum a^{1+x} b^{1-x}) \tag{3}$$

also lies between $F$ and $G$ where $x$ is a real number between 0 and 1. Indeed, when $x = 0$, $R = G$, and as $x$ increases, $R$ increases and reaches the value $F$ when $x = 1$.

## 2. CONVEX FUNCTIONS

Daykin and I have considered these questions further.[4,5] We have shown that certain inequalities may be generalized by the construction of suitable convex functions. [By a convex function is meant a function $f(x)$ satisfying $f((x_1 + x_2)/2) < \frac{1}{2}\{f(x_1) + f(x_2)\}$ for every pair of unequal values $x_1$ and $x_2$. If $f(x)$ is continuous and is differentiable till second order, then the condition for $f(x)$ to be convex is equivalent to $d^2f/dx^2 > 0$ at every point of the range in question.]

The Callebant function $R$, given in (3) above, is a convex function of $x$, with minimum at $x = 0$, except if the sets $(a)$ and $(b)$ are proportional in which case $R$ is a constant.

We have proved[4] the following theorem:

*Theorem I:* The function

$$S(x) = (\sum a^{1+\alpha x} b^{1+\beta x} c^{1+\gamma x})(\sum a^{1+\gamma x} b^{1+\alpha x} c^{1+\beta x})(\sum a^{1+\beta x} b^{1+\gamma x} c^{1+\alpha x})$$

$$(4)$$

where $\alpha + \beta + \gamma = 0$ is a convex function of $x$, with minimum at $x = 0$, except if the sets $(a), (b)$, and $(c)$ are proportional in which case $S$ is a constant. The function has been generalized to $n$ sets $(a), (b), \ldots$ (but I will not quote the long expression here).

Again in relation to Hölder's inequality, we construct[5] the function

$$H(x) = \left\{ \sum d \left( \frac{a^p}{d} \right)^x \right\}^{1/p} \left\{ \sum d \left( \frac{b^q}{d} \right)^x \right\}^{1/q} \tag{5}$$

where $d_k = a_k b_k, k = 1, 2, \ldots, n$; and prove the theorems:

*Theorem II:*

1. If $(1/p + 1/q) < 1$, $H(x)$ is a convex function (except if all the $a$'s are equal, all the $b$'s are equal, and $a_i b_i$ is unity, in which case $H(x)$ is a constant).

2. If $(1/p + 1/q) = 1$, $H(x)$ is a convex function with minimum at $x = 0$ (except if the sets $(a^p)$ and $(b^q)$ are proportional in which case $H(x)$ is a constant).

3. If $(1/p + 1/q) > 1$ and sufficiently large, $H(x)$ is concave (except if all the $a$'s are equal, all the $b$'s are equal and $a_i b_i$ is unity in which case $H(x)$ is constant).

*Theorem III:*

1. If $(1/p + 1/q) < 1$ then for all $a_i b_i > 0$

$$2(\sum ab)^{1/p+1/q} \leq (\sum a^p)^{1/p}(\sum b^q)^{1/q} + (\sum a^{2-p_{bi}})^{1/p}(\sum a^2 b^{2-q})^{1/q} \tag{6}$$

2. If $(1/p + 1/q) = 1$, then for all $a_i b_i > 0$,

$$\sum ab \leq (\sum a^p)^{1/p} (\sum b^q)^{1/q} \tag{7}$$

which is Hölder's inequality.[1]

3. If $(1/p + 1/q) < 1$, and if every $a_i$ and $b_i$ is less than unity or if $P = {}_i\prod_j (a_i a_j b_i b_j)^{a^i a_j b^i b_j} < 1$, then

$$(\sum ab)^{1/p+1/q} \leq (\sum a^{2-p_{b^i}})^{1/p} (\sum a^2 b^{2-q})^{1/q} \tag{8}$$

4. If $(1/p + 1/q) < 1$ and if every $a_i$ and $b_i$ is greater than unity or if $P > 1$,

$$(\sum ab)^{1/p+1/q} \leq (\sum a^p)^{1/p} (\sum b^q)^{1/q} \tag{9}$$

Again the function

$$M(x) = \{\sum(\alpha + \beta x)^p\}^{1/p} \tag{10}$$

has been shown to be related to Minkowski's inequality, and the following theorem proved:

*Theorem IV:* For values of $x$ such that all the $\alpha_i + \beta_i x \geq 0$, the function $M(x)$ is convex or concave according as $p > 1$ or $p < 1$.

From this theorem, inequalities of interest have been derived.


## 3. APPLICATIONS

As examples of further applications, we may consider results obtained from the above by passing from summation to integration. Thus, from the expression (4) above we are led to the following theorem:

*Theorem V:* If $f(\tau), g(\tau)$, and $h(\tau)$ are non-negative real functions defined for $a \leq \tau \leq b$, the function

$$I(x) = \left(\int_a^b f^{1+\alpha x} g^{1+\beta x} h^{1+\gamma x} d\tau\right)\left(\int_a^b f^{1+\gamma x} g^{1+\alpha x} h^{1+\beta x} d\tau\right)\left(\int_a^b f^{1+\beta x} g^{1+\gamma x} h^{1+\alpha x} d\tau\right) \tag{11}$$

is convex with minimum at $x = 0$ (provided the integrals exist and unless $f, g$, and $h$ are proportional almost everywhere in $a \leq \tau \leq b$). By taking suitable functions $f, g$, and $h$, interesting inequalities may be obtained.

I refer to some of those involving $\Gamma$ functions, obtained by these methods:

(i) 
$$T(x) = \frac{\Gamma(\sigma + \lambda x)\Gamma(\sigma + \mu x)\Gamma(\sigma + \nu x)}{(\sigma' + \lambda' x)^{\sigma + \lambda x}(\sigma' + \mu' x)^{\sigma + \mu x}(\sigma' + \nu' x)^{\sigma + \nu x}} \quad (12)$$

where $\sigma, \lambda, \mu, \nu; \sigma', \lambda', \mu', \nu'$ are real numbers with $\lambda + \mu + \nu = \lambda' + \mu' + \nu' = 0$ is convex with minimum at $x = 0$ when the six numbers in brackets are all positive.

(ii) If $a_1, a_2, \ldots, a_n$ are positive real numbers with average $A = (a_1 + \cdots + a_n)/n$,

$$\frac{\Gamma(a_1)\Gamma(a_2) \cdots \Gamma(a_n)}{\{\Gamma(A)\}^n} > \frac{a_1^{a_1} a_2^{a_2} \cdots a_n^{a_n}}{(A^A)^n} \quad (13)$$

(iii) If $(1/p + 1/q) = 1$,

$$\frac{\{\Gamma(1 + \alpha p)\}^{1/p}\{\Gamma(1 + \beta q)\}^{1/q}}{\Gamma(1 + \alpha p + \beta q)} \geq \frac{(1 + \alpha p)^{(1 + \alpha p)/p}(1 + \beta q)^{(1 + \beta q)/q}}{(1 + \alpha + \beta)^{1 + \alpha + \beta}} \quad (14)$$

where the term in each bracket is positive.

$$\frac{\Gamma(a - x)\Gamma(a + x)}{\Gamma(a - y)\Gamma(a + y)} > \frac{(a - x)^{a-x}(a + x)^{a+x}}{(a - y)^{a-y}(a + y)^{a+y}} \quad a > x > y \geq 0 \quad (15)$$

A particular case of this may be noted in the following form:

$$\frac{\Gamma(a)\Gamma(b)}{\{\Gamma[(a + b)/2]\}^2} \geq \frac{a^a b^b}{[(a + b)/2]^{a+b}} \quad a, b > 0 \quad (16)$$

## REFERENCES

1. G. H. Hardy, J. E. Littlewood, and G. Polya, "Inequalities," University of California Press, 1952.
2. E. A. Milne, "Note on Rosseland's integral for the stellar absorption coefficient," *Monthly Notices Roy. Astron. Soc.* **85**: 979 (1925).
3. D. K. Callebant, "Generalization of the Cauchy-Schwarz inequality," *J. Math. Anal. Appl.* **12**: (1965).
4. C. J. Eliezer and D. E. Daykin, "Generalizations and applications of Cauchy-Schwarz inequality," *Q. J. Math (Oxford)* **18**: 357 (1967).
5. D. E. Daykin and C. J. Eliezer," Generalization of Hölder's and Minkowski's inequalities," to appear in *Proc. Camb. Phil. Soc.*

# On Harmonic Differential Forms in a General Manifold

NIRMALA PRAKASH

*DELHI UNIVERSITY*
*Delhi, India*

---

During the past fifteen years the theory of form-calculus has been mainly responsible for the development of differential geometry. The roots of the subject are to be found in E. Cartan's famous paper of 1934 on Finster spaces, and later on the subject flourished at the hands of S. Chern, J. A. Schouten, K. Kodiara, S. Bochner, A. Hilt, S. Hseiung, and several others.

With the establishing of De Rahm's isomorphism theorem[9] which states that if $M$ is a compact orientable Riemannian manifold then, the number of linearly independent harmonic forms of degree $p$ is equal to the $p$-th betti-number of $M$,‡ the study of differential geometry through form calculus became still more interesting, since $p$th betti numbers which are topological invariants of a manifold could be obtained by finding the linearly independent harmonic forms of $M$. Particularly the fact that $p$-th order betti-numbers of a manifold are zero could be easily ascertained by showing that there exist no nontrivial harmonic-forms of order $p$ on the manifold. Several authors, namely, S. Bochner, K. Yano, R. Couty, W. V. D. Hodge, A. Lichnerowiz, I. Mogi, De Rahm, S. I.

---

‡This theorem is also known to hold for paracompact spaces
Latin indices $i, j, k \ldots$ take the values 1 to $n$.
Greek indices $\alpha, \beta, \gamma \ldots$ stand for members of $J_+$.

Goldberg, and others studied the problems of existence of such vectors and tensor fields on compact Riemmannian manifold. The present author while working on those lines was motivated to obtain similar results for manifolds which are not necessarily compact.

Since Green's theorem has been the main tool in establishing these results, we shall show in Section 1 Green's theorem can be proved for paracompact manifolds modeled on Hilbert space.[7,8] In Section 2, its application is illustrated by proving a result in such a manifold.

To make the account self-contained some of the basic concepts have been given in detail.

## 1. GREEN'S THEOREM

### Preliminaries

A hermitian form on a real (complex) vector space $E$ is a mapping $\psi$ of $E \times E$ into $R(G)$ which has the following properties[‡]

$$\psi(x + x', y) = \psi(x, y) + \psi(x', y) \tag{1}$$

$$\psi(x, y + y') = \psi(x, y) + \psi(x, y') \tag{2}$$

$$\psi(\lambda x, y) = \lambda \psi(x, y) \tag{3}$$

$$\psi(x, y) = \overline{\psi(y, x)} \tag{4}$$

A hermitian form $\psi$ is said to be positive if $\psi(x, x) \geq 0$ for any $x \in E$ and it is said to be nondegenerate if

$$\psi(x, x) = 0 \leftrightarrow x = 0$$

It can be easily verified that a nondegenerate positive hermitian form one $E$ defines a norm thereon. A space $E$ having a nondegenerate positive hermitian form is called a prehilbert space and a prehilbert space which is complete is called a *Hilbert Space*.

### Differentiable Manifold on Hilbert Space

A topological space $S$ is said to be paracompact if every open cover $A$ of $S$ admits of a subcover $B$ such that[1]

---

‡$R$, $G$, $H$ stand, respectively, for real, complex, and Hilbert spaces, whereas $E$ and $S$ stand for an arbitrary space.

1. It is a refinement of $A$, i.e., every member of $B$ is contained in some memer of $A$.
2. It is locally finite, i.e., every point of $S$ has a neighbourhood intersecting finitely many members of $B$.

A topological manifold $M$ is a Hausdorff space satisfying the second axiom of countability.

A $C^r$-differentiable manifold $M$ modeled on Hilbert space is a topological manifold together with a differentiable structure $D$ of class $C^r$ is a collection of coordinate neighborhoods $(u_\alpha, h_\alpha)$ covering $M$ and satisfying the following two conditions:[2]

1. For any two pairs $(u_\alpha, h_\alpha), (u_\beta, h_\beta) \in D$ the homeomorphic map $h_\alpha h_\beta^{-1} : h_\beta(u_\alpha \cap u_\beta) \to$ to open subsets of Hilbert spaces is $C^r$-differentiable.
2. $D$ is maximal with respect to the above property.

A manifold $M$ is said to be orientable if there exists at least one differentiable structure $D$ on it in which for every pair of co-ordinate neighborhoods with non-empty intersection the mapping $h_\alpha h_\beta^{-1}$ is sense-preserving.

The differentiable manifold $M$ throughout this note is orientable paracompact, and is modeled on a Hilbert space.[7]

Let $\mathscr{F}$ denote the collection of all non-negative functions on $M$, and let $f$ be an arbitrary member of $\mathscr{F}$ then the closure of the set of points $m$ in $M$ for which $f(m) > 0$ is called the carrier of $f$. A family of differentiable functions on any differentiable manifold $M$ is a partition of unity if it satisfies the following three conditions:[4]

1. Each member of $\mathscr{F}$ is non-negative
2. The family of carriers of members of $\mathscr{F}$ forms a locally finite covering of $M$
3. For every $m \in M$ and $f_\alpha \in \mathscr{F}$, $\sum\limits_{\alpha=1}^{\infty} f_\alpha(m) = 1$.

It is known that on a differentiable maniford $M$ one can always find a function $h_\alpha$ such that

$$h_\alpha(m) = 1 \quad \text{for} \quad \|m\| \leq \nu_1$$
$$h_\alpha(m) = 0 \quad \text{for} \quad \|m\| \geq \nu_2$$

Where $\nu_1$ and $\nu_2$ are real numbers satisfying $0 < \nu_1 < \nu_2$.

Since $M$ is paracompact, $\sum\limits_{\alpha=1}^{\infty} h_\alpha(m)$ is meaningful and the collection $\{f_\alpha(m)\}$ of functions defined as $f_\alpha(m) = h_\alpha(m)/\sum\limits_{\alpha=1}^{\infty} h_\alpha(m)$ forms a partition of unity.

It is also known that if $M$ is assumed to be connected a Riemannian metric can be trivially defined on $M$.

We now state and prove the theorem locally and observe that condition (1) of differentiable structure $D$ ensures its global extension. (For convenience, the dimension of $M$ is taken to be $n$.)

*Theorem:* In a paracompact orientable manifold $M$, for any arbitrary vector field $\lambda^i$ we have

$$\int_M \lambda^i_{;i}\, dv = 0‡$$

where $dv = \sqrt{g}\, dx^1 \wedge dx^2 \wedge \cdots \wedge dx^n$ ($g = |g_{ij}|$ being the determinant of Riemannian metric tensor).

*Proof:* Let $U = \{u_\alpha\}$ denote the cover of $M$ and a family $\mathscr{F} = \{f_\alpha\}$ be a partition of unity, and let carrier $f_\alpha$ be denoted by $\tilde{w}_\alpha$, then since $W = \{\tilde{w}_\alpha\}$ is a refinement of $U$, for each $\tilde{w}_\alpha$ there is a coordinate neighborhood $u_\alpha$ say which contains $\tilde{w}_\alpha$.

Define

$$\lambda^i_\alpha = f_\alpha \lambda^i \tag{5}$$

Clearly

$$\sum_{\alpha=1}^\infty \lambda^i_\alpha = \sum_{\alpha=1}^\infty f_\alpha \lambda^i = \lambda^i$$

Consequently,

$$\lambda^i_{;i} = \sum_{\alpha=1}^\infty \lambda^i_{\alpha;i}$$

Now

$$\int_M \lambda^i_{\alpha;i}\, dv = \int_{M \sim \tilde{w}_\alpha} \lambda^i_{\alpha;i}\, dv + \int_{\tilde{w}_\alpha} \lambda^i_{\alpha;i}\, dv$$
$$= I_1 + I_2$$

In view of relation (5), $\lambda^i_\alpha$ vanishes throughout $M \sim \tilde{w}_\alpha$, consequently $I_1$ is zero.

Also, $\tilde{w}_\alpha$ is a closed subregion of an orientable manifold. Therefore integral taken around any contour in this region vanishes; hence, $I_2$ is zero.

Thus,

$$\int_M \lambda^i_{\alpha;i}\, dv = 0$$

---

‡ ; or $\nabla$ followed by a subscript stands for covariant derivative.

Since

$$\int \lambda_{\alpha;i}^i \, dv = \sum_{\alpha=1}^{\infty} \int \lambda_{\alpha;i}^i \, dv$$

We have the required result.

## 2. HARMONIC FORMS AND BETTI-NUMBERS

A differential form of degree $p$ on a manifold $M$ is the invariant $(1/p!)\omega_{i_1 i_2 \cdots i_p} \, dx^{i_1} \wedge dx^{i_2} \cdots \wedge dx^{i_p}$ where $\omega_{i_1 \cdots i_p} = \omega_{I(p)}$ is a skew symmetric tensor of order $p$, $I(p) = (i_1 \cdots i_p)$ being the ordered subset of the set $(1, 2, 3, \ldots, n)$.

Exterior differentiation of $\omega = \omega_j \, dx^j$ is defined as

$$d\omega = \tfrac{1}{2}(\partial_j \omega_i - \partial_i \omega_j) \, dx^j \wedge dx^i$$
$$= \tfrac{1}{2}(\nabla_j \omega_i - \nabla_i \omega_j) \, dx^j \wedge dx^i$$

But $\nabla_j \omega_i - \nabla_i \omega_j = \text{Rot } \omega$, hence we sometimes denote Rot $\omega$ by $d\omega$. Again Div. $\omega \stackrel{\text{def}}{=} \nabla_i \omega^i$ and in case $\omega_i = \nabla_i f$ ($f$ being a scalar) then $g^{ij}\nabla_j \omega_i = \text{Div } \omega$. Alternatively Div. $\omega$ is called the co-differential of the differential form $\omega = \omega_i \, dx^i$ and is denoted by $\delta\omega$.

The (exterior) differential and co-differential of a $p$-form $w$ are, respectively, defined by

$$dw = \frac{1}{(p+1)!} (\nabla_j w_{i_1 i_2 \cdots i_p} - \nabla_{i_1} w_{j i_2 \cdots i_p} - \cdots - \nabla_{i_p} w_{i_1 \cdots i_{p-1} j})$$
$$\wedge \, dx^j \wedge dx^{i_1} \cdots \wedge dx^{i_p}$$
$$\delta w = (g^{ij}\nabla_j w_{i i_2 \cdots i_p}) \, dx^{i_2} \wedge \cdots dx^{i_p}$$

It is obvious that operators $d$ and $\delta$ map a $p$-form to a $(p + 1)$ and $(p - 1)$ form, respectively. They also satisfy the following:

$$ddw = 0 \qquad \delta\delta w = 0 \tag{3}$$

and

$$\delta w = (-1)^{p(n-p)} * d * w \tag{4}$$

where $*w$ is the form, dual of $w$ and is defined as

$$\frac{1}{p!} \sqrt{g} \; \epsilon_{i_1 i_2 \cdots i_{p+1} \cdots i_n} w^{i_1 \cdots i_p} \, dx^{i_{p+1}} \wedge dx^{i_{p+2}} \cdots \wedge dx^{i_n}$$

A class of forms whose $d$-image ($\delta$-image) is zero is called closed (co-closed).

A form which is closed and co-closed is called a harmonic form, and the corresponding tensor a harmonic tensor, thus a harmonic form is characterized by $dw = 0$ and $\delta w = 0$ or equivalently by $w_{[i_1 i_2 \cdots i_p; j]} = 0$ and $g^{ij} \nabla_i w_{j; i_2 i_3 \cdots i_p} = 0$.

A form $w$ which is a $d$-image ($\delta$-image) of some other form is called exact (co-exact).

In view of equation (3) each exact (co-exact) form is obviously closed (co-closed).

Now the totality of closed $p$-forms on a manifold forms a group with respect to addition of forms as the group operation and the collection of exact $p$-forms is a subgroup thereof. Similarly the collection of co-exact $p$-forms is a subgroup of the group formed by co-closed $p$-forms.

The group of equivalence classes of closed $p$-forms modulo the exact $p$-forms is denoted by $D^p(M)$ and dim $D^p(M)$ is equal to the number of linearly independent harmonic tensors of order $p$.

On the other hand $b_p(M)$ the $p$-th betti-number of manifold $M$ is the dim of cohomology group $H^p(M)$ of $M$, consequently De Rahm's theorem asserts:

$$H^p(M) \cong D^p(M)$$

$M$ being a paracompact manifold.

Let $w = \xi_{i_1 i_2 \cdots i_p} dx^{i_1} \wedge \cdots \wedge dx^{i_p}$ be an arbitrary form on $M$ and let us consider the divergence

$$\nabla_j \{ \xi^i_{I(p-1)} (\nabla_i \xi^{jI(p-1)}) - \xi^j_{I(p-1)} (\nabla_i \xi^{iI(p-1)}) \}$$

We have

$$\nabla_j \{ \xi^i_{I(p-1)} (\nabla_i \xi^{jI(p-1)}) - \xi^j_{I(p-1)} (\nabla_i \xi^{iI(p-1)}) \}$$
$$= F(\xi_{I(p)}) + (\nabla^j \xi^{iI(p-1)})(\nabla_i \xi_{jI(p-1)}) - (\nabla_j \xi^i_{I(p-1)})(\nabla_i \xi^{iI(p-1)}) \qquad (1)$$

where $F(\xi_{I(p)})$ stands for

$$R_{ji} \xi^j_{I(p-1)} \xi^{iI(p-1)} + \frac{p-1}{2} R_{ijkl} \xi^{ij}_{I(p-2)} \xi^{klI(p-2)};$$

$R_{ijkl}$ and $R_{ij}$ being curvature tensor and Ricci tensors, respectively.

Applying Green's theorem to the vector

$$\{ \xi^i_{I(p-1)} (\nabla_i \xi^{jI(p-1)}) - \xi^j_{I(p-1)} (\nabla_i \xi^{iI(p-1)}) \}$$

we have

$$\int_M \{ F(\xi_{I(p)}) + (\nabla^j \xi^{iI(p-1)})(\nabla_i \xi_{jI(p-1)}) - (\nabla_j \xi^i_{I(p-1)})(\nabla_i \xi^{iI(p-1)}) \} \, dv = 0$$

and this can be written in the form

$$\int_M \{2pF(\xi_{I(p)}) + 2(\nabla^j \xi^{iI(p-1)})(\nabla_j \xi_{iI(p-1)}) - 2(p+1)(\nabla^{[j}\xi^{iI(p-1)]})$$
$$\times (\nabla_{[j}\xi_{iI(p-1)]}) - 2p(\nabla_j \xi^j_{I(p-1)})(\nabla_i \xi^{iI(p-1)})\} \, dv = 0$$

Since $\xi_{I(p)}$ is skew symmetric, above equation can further be reduced to the form

$$\ddagger \int_M \{k^{(p)}_{ijkl}\xi^{klI(p-2)}\xi^{ij}_{I(p-2)} + G_{ijklmn}(\nabla^k \xi^{ijI(p-2)}) \times (\nabla^n \xi^{lm}_{I(p-2)})$$
$$- 2(p+1)(\nabla^{[j}\xi^{iI(p-1)]})(\nabla_{[j}\xi_{iI(p-1)]})$$
$$- 2p(\nabla_j \xi^j_{I(p-1)})(\nabla_i \xi^{iI(p-1)})\} \, dv = 0 \qquad (2)$$

If we choose $w$ to be a harmonic $p$-form, then (2) takes the form

$$\int_M \{k^{(p)}_{ijkl}\xi^{klI(p-2)}\xi^{ij}_{I(p-2)} + G_{ijklmn}(\nabla^k \xi^{ijI(p-2)}) \times (\nabla^n \xi^{lm}_{I(p-2)})\} \, dv = 0$$

Consequently we have the following result:

*Result*: If in a paracompact orientable manifold $M$ modeled on Hilbert space the matrix $N$

$$N = \left\| \begin{matrix} k^{(p)}_{ijkl} & 0 \\ 0 & G_{ijklmn} \end{matrix} \right\|$$

defines a quadratic form in the variables $\xi^{rs}$ and $\eta^{rst} = \nabla^t \xi^{rs}$ then there exists no nontrivial harmonic form of order $p$ in $M$ hence the $p$-th betti-number of such a manifold is zero.

It is to be remarked that Green's theorem[5] can also be proved for noncompact affine manifolds satisfying the property $(P)$ (i.e., every open cover of $M$ admits a finite subcollection whose closures cover $M$) by restricting the affine connection suitably, and thus many results of orientable compact metric manifolds can be extended to orientable noncompact affine manifolds.

## REFERENCES

1. J. R. Munkres, "Elementary Differential Topology," Princeton University Press, Princeton, 1963.

---

$$\ddagger k^{(p)}_{ijkl} = \tfrac{1}{2}p(R_{ik}g_{lj} - R_{jk}g_{li} - R_{il}g_{kj} - R_{jl}g_{ki})$$
$$- \tfrac{1}{2}p(p-1)(R_{iklj} - R_{jkli} - R_{ilkj} - R_{jlki})$$

and

$$G_{ijklmn} = (g_{il}g_{jm} - g_{im}g_{jl})g_{kn}$$

2. L. Auslander and R. E. Mackenzie, "Introduction to Differentiable Manifolds," McGraw-Hill Book Co., New York, 1963.

3. S. T. Hu, "Elements of General Topology," Holden-Day-Inc., 1965.

4. S. I. Goldberg, "Curvature and Homology," Academic Press, New York, 1962.

5. N. Prakash, "Green's Theorem in Noncompact Affine-Manifolds," (to appear in National Instt. Sc. India).

6. J. Dieudonne, "Une Generalisation des espace compacts," *J. Math. Pures Appl.* **23**: (1944).

7. S. Lang, "Introduction to Differentiable Manifolds," Interscience Publication, New York, (1962–63).

8. J. Dieudonne, "Foundations of Modern Analysis," Academic Press, New York, 1960.

9. G. De. Rham, "Varietes differentiables," Actualités scientifiques et industrielles, No. 1222, Herman, Paris, 1955.

10. K. Yano, "Differential Geometry of Complex and Almost Complex Spaces, Pergamon Press, New York, 1965.

11. N. Prakash, "Green's theorem in paracompact manifolds modeled on Hilbert spaces," to appear in Proceedings of Indian Academy of Sciences.

# Developments in the Theory of Univalent Functions

K. S. Padmanabhan

*ANNAMALAI UNIVERSITY*
*Annamalainagar, India*

---

A univalent function in a domain $D$ is characterized by the property that it takes no value more than once in the domain and that, consequently, it maps the domain $D$ onto a Schlicht domain, that is, one which is not self-overlapping and contains no branch points. In investigating the properties of analytic functions univalent in a domain $D$ which is simply connected, one usually confines oneself to the case of the unit disc, for, by the Riemann mapping theorem, any simply connected domain with at least two boundary points can be mapped onto the unit disc, and any univalent function in $D$ is associated with a corresponding univalent function in the unit disc. If $f(z)$ is regular and univalent in the unit disc, so also is $\{f(z) - f(0)\}/f'(0)$, and this enables us to use the normalization $f(0) = 0, f'(0) = 1$ and study the class $S$ of normalized regular functions having the Taylor expansion

$$f(z) = z + a_2 z^2 + \cdots \qquad |z| < 1 \qquad (1)$$

We may observe that $f'(0)$ does not vanish. If it did, $f(z)$ could not be univalent. There also exist functions which are univalent but not regular in the unit disc. For example, if $f(z) \in S, \{af(z) + b\}/\{cf(z) + d\}, ad - bc \neq 0$ is obviously univalent in $|z| < |$ but has a simple pole if $-d/c$ is one of the values assumed by $f(z)$ in $|z| < 1$. But a univalent function cannot possess any singularities other than a simple pole. If it did, the value $\infty$ would be taken more than once. All such functions may be normalized by the requirement that the pole may be located at $z = 0$ and the residue is unity. The class of such functions will be denoted by the symbol $P$. A paper by P. Koebe[14] in 1907 may be rightly regarded as the starting point in the investigation of univalent functions. Therein, Koebe established the existence of a constant $K$ (Koebe's constant) such that the boundary of the map of the unit disc, by any function $w = f(z) \in S$ is always at a distance not less than $K$ from $w = 0$. He also presented the theorem of distortion bearing his name, in which he proved the existence of bounds for $|f'(z)|$, these bounds depending only on $|z|$. Koebe's results soon attracted the attention of others, and Gronwall[9] was the first to give the so-called area theorem which asserts that if $g(z) = 1/z + b_0 + b_1 z + \cdots$ belongs to $P$, then $\sum n|b_n|^2 \leq 1$. This inequality is the analytic expression of the geometrical fact that the area of the domain left uncovered by the image of $|z| = r < 1$ by $w = g(z)$ is positive. Bieberbach[2] and Faber[7] rediscovered the area principle in 1916 and used it to obtain the precise bounds, for Koebe's constant $K$, for $|f'(z)|, |f(z)|$ and $a_2$. It was found that $K = 1/4$ and that

$$\frac{(1 - |z|)}{(1 + |z|)^3} \leq |f'(z)| \leq \frac{(1 + |z|)}{(1 - |z|)^3} \tag{2}$$

The validity of the upper bound for all $|z| < 1$ gives the precise inequality $|a_2| \leq 2$. Equality occurs in equation (2) only for the Koebe's function

$$K(z) = \frac{z}{(1 + e^{i\varphi} z)^2} \qquad \varphi \text{ real} \tag{3}$$

which maps the unit disc onto the plane minus the half-line issuing from the point $w = 1/4\, e^{-i\varphi}$ to $w = \infty$.

The extremal character of the function $k(z)$ in connection with the bounds for $|f'(z)|, |f(z)|$ and $|a_2|$ gave rise to the conjecture (Bieberbach's conjecture) that $|a_n| \leq n, n = 2, 3, \ldots,$ for the function $f(z) \in S$. As yet, this famous conjecture has neither been proved nor disproved, although a lot of information about univalent functions and a variety of new tools of research have sprung up from all such efforts. A number of subclasses of $S$ have been introduced and their properties widely investigated. It is our intention to study these subclasses in some detail. The first subclass of $S$ to be studied was that of convex functions.[32] These are the functions which map the unit disc onto convex domains. These functions have also been investigated subsequently by Gronwall,[10] Lowner[21] and others. A domain $D$ is said to be convex if the line segment joining any two points of $D$ lies wholly in $D$. It is then easy to deduce that the center of gravity $(\omega_1 + \omega_2 + \cdots + \omega_n)/n$ of $n$ points $\omega_1, \omega_2, \ldots, \omega_n$ in $D$ also lies in $D$. If $f(z) \in S$ maps $|z| < 1$ onto a covex domain, it maps all circles $|z| = \rho < 1$ onto convex curves (Ref. 22, p. 222). Another important subclass of $S$ consists of the starlike functions first treated by Alexander[1] and later by Nevanlinna[23] and others. These functions map $|z| < 1$ onto a domain star shaped with respect to the origin. A domain $D$ containing the point $\omega = 0$ is said to be starlike (or star shaped) with respect to $\omega = 0$ if the line segment joining $\omega = 0$ to any other point of $D$ lies wholly in $D$. If $f(z) \in S$ maps $|z| < 1$ onto a starlike domain $D$, then every disc $|z| < \rho < 1$ is also mapped onto a domain of the same type[22] (p. 220). From this we can deduce that a necessary and sufficient condition for $f(z)$ to map the unit disc onto a starlike domain is

$$\mathrm{Re}\left\{\frac{zf'(z)}{f(z)}\right\} > 0 \qquad |z| < 1 \tag{4}$$

It is also easily seen that $f(z)$ maps $|z| < 1$ onto a convex domain if and only if $zf'(z)$ maps $|z| < 1$ onto a starlike domain, that is, if and only if

$$\mathrm{Re}\left\{\frac{1 + zf''(z)}{f'(z)}\right\} > 0 \qquad |z| < 1 \tag{5}$$

The inequalities $|a_n| \leq n$ are true for the starlike functions and are sharp, since $z/(1 - z)^2 = z + 2z^2 + \cdots + nz^n + \cdots$ is starlike in $|z| < 1$. The estimate of $|a_n|$ can be lowered for the convex functions which form a subclass of the starlike functions. Indeed, $|a_n| \leq 1$ for a convex function as it is readily seen from the fact that $f$ is convex if and only if $zf'$ is starlike in $|z| < 1$. Here again the estimate is sharp as is shown by the function $f(z) = z/(1 - z)$ which is convex in $|z| < 1$. Kaplan[13] introduced the class of close-to-convex univalent functions in the unit disc. A function $f(z)$ of this class is characterized by the property that there exists an analytic function $\varphi(z)$ univalent and convex in $|z| < 1$ such that

$$\mathrm{Re}\left\{\frac{f'(z)}{\varphi'(z)}\right\} > 0 \qquad |z| < 1 \tag{6}$$

Since $F(z) = z\varphi'(z)$ is starlike in $|z| < 1$ whenever $\varphi(z)$ is convex, condition (6) is equivalent to

$$\text{Re}\left\{\frac{zf'(z)}{F(z)}\right\} > 0 \qquad |z| < 1 \tag{7}$$

where $F(z)$ is starlike in $|z| < 1$. Kaplan has proved that every close-to-convex function is also univalent in the unit disc. By specializing $\varphi(z)$, one obtains other subclasses. For $\varphi(z) = z$ one gets the class of functions whose derivatives have positive real part in the unit disc. This class has been widely investigated by several authors. J. Wolff[35] was the first to prove that if $f(z)$ is regular in a half plane $D$ and Re $f'(z) = 0$ there, then $f(z)$ is univalent in $D$. Noshiro[24] (p. 151) and Warschawski[34] independently demonstrated that Re $f'(z) > 0$ implies the univalence of $f(z)$ in any convex domain. On the other hand, Herzog and Piranian[12] proved that if a domain $D$ is such that the statement Re $f'(z) > 0$ for $z \in D$ implies that $f(z)$ is univalent in $D$, then $D$ is not far from being convex. Umezawa[33] obtained certain sufficient conditions for univalence of a function $f(z)$ in a simply connected domain. Let $w = f(z)$ be regular in a simply connected closed region $D_z$ whose boundary $\Gamma_z$ consists of a regular curve and let $f'(z) \neq 0$ on $\Gamma_z$. If one of the following conditions hold:

1. For arbitrary arcs $C_z$ on $\Gamma_z$, $\int_{C_z} d \arg df(z) > -\pi$, and $\int_{\Gamma_z} d \arg df(z) = 2\pi$

2. For arbitrary arcs $C_z$ on $\Gamma_z$, $\int_{C_z} d \arg df(z) < 3\pi$, then $f(z)$ is univalent in $D_z$.

As Reade[27] has remarked, Umezawa's criteria for univalence and Kaplan's criteria for a function to be close-to-convex are essentially the same. Employing Umezawa's criteria, Reade[28] obtained the following theorem:

*Theorem:* Let $\varphi$ be fixed and $0 \leq \varphi \leq \pi$. Let $D$ be a domain in which it is possible to join each pair of distinct points $z_1, z_3$ by a pair of straight line segments $\overrightarrow{z_1 z_2}, \overrightarrow{z_2 z_3}$ in $D$ such that

$$\left|\text{Arg}\,\frac{z_3 - z_2}{z_2 - z_1}\right| \leq \varphi \tag{8}$$

$\varphi$ being independent of $z_1$ and $z_3$ in $D$. Then if $f(z)$, is analytic in $D$ and if for some $0 \leq \alpha \leq 2\pi$, $\alpha \leq \text{Arg}\,f'(z) \leq \pi - \varphi + \alpha$, for $z \in D$, then $f(z)$ is univalent in $D$.

Similar conditions for univalence of $f(z)$ have been obtained by Cowling and Royster.[6] A typical theorem they have proved in this connection is as follows:

*Theorem:* Let $D$ be a domain any two points of which may be joined by the arc (lying in $D$) of an ellipse $z = z_0 + e^{i\beta}(a \cos t + ib \sin t)$, for which $t_1 \leq t \leq t_2, 0 \leq t_1, t_2 \leq \pi/2$ and where $a, b > 0$, $\beta$ real ($a, b, \beta$ depend on the two points to be connected.) Then if $f(z)$ is analytic in $D$ and if for some $0 \leq \alpha \leq 2\pi$

$$\alpha < \text{Arg} f'(z) < \alpha + \frac{\pi}{2}$$

for all $z \in D$, then $f(z)$ is univalent in $D$.

Returning to the close-to-convex functions, it is interesting to note that Kaplan has characterized these functions intrinsically, without reference to an auxiliary convex function. A necessary and sufficient condition that a function $f(z)$, analytic and with non-vanishing derivative for $|z| < 1$, be close-to-convex is that

$$\int_{\theta_1}^{\theta_2} \text{Re}\left[\frac{1 + re^{i\theta} f''(re^{i\theta})}{f'(re^{i\theta})}\right] d\theta > -\pi \tag{9}$$

for $\theta_1 < \theta_2, 0 < r < 1$. The geometric interpretation of (9) is that $w = f(z)$ maps each circle $|z| = r < 1$ onto a simple closed curve whose tangent rotates, as $\theta$ increases, either in the counterlockwise direction, or clockwise in such a way that it never turns back on itself so much as to completely reverse its direction. For $f(z) = z + a_2 z^2 + \cdots + a_n z^n + \ldots$, close-to-convex in $|z| < 1$ Reade[26] has shown that $|a_n| \leq n, n = 2, 3, \ldots$ thus, extending the validity of Bieberbach's conjecture to a wider class than was previously considered.[29] Recently, Hayman[11] proved that $||a_{n+1}| - |a_n|| < A|a_1|$, for some absolute constant $A$, where $f(z) = \sum_1^\infty a_n z^n$ is univalent in $|z| < 1$. If $f(z)$ is, however, close-to-convex in $|z| < 1$, according to a result of Pommerenke,[25]

$$||a_{n+1}| - |a_n|| < \frac{3}{4} e^2 |a_1| \tag{10}$$

Let $M(r) = \max_{|z|=r} |f(z)|$. Then

$$M(r) = 0((1 - r)^{-\alpha}) \longrightarrow a_n = 0(n^{\alpha-1}) \tag{11}$$

provided $\alpha > 1/2$, as Littlewood and Paley[19] have shown for the general case where $f(z)$ is any univalent function. The implication (11) does not hold for small $\alpha$. However, if $f(z)$ is close-to-covex in $|z| < 1$, Clunie and Pommerenke have demonstrated that (11) holds for $\alpha \geq 0$.[5]

The close-to-convex functions include several familiar classes of univalent functions. The starlike functions form a subclass with $F(z) = f(z)$ in (7) and the convex functions themselves are clearly included in the class of close-to-convex functions. Another important sub-class is formed by the functions that are convex in one direction introduced by Robertson.[29] For such functions the intersection of the image region with each straight line of a certain fixed direction is either empty or one interval. A function $f(z)$ regular in $|z| < 1$ is convex in one direction if and only if there is a starlike function $F(z) = b_1 z + b_2 z^2 + \cdots$ in $(z) < 1$, which is of the form $F(z) = b_1 z / \{(1 - e^{-i\theta_1} z)(1 - e^{-i\theta_2} z)\}$ with real $\theta_1$ and $\theta_2$ and complex $b_1$ such that Re $\{zf'(z)/F(z)\} > 0$, in $|z| < 1$.

The priciple of subordination has also been exploited by several writers to derive new results in the theory of univalent functions. If the analytic function $f(z) = \sum\limits_{n=1}^{\infty} a_n z^n (|z| < 1)$ is merely regular (not necessarily univalent) in the unit disc and assumes there no value omitted by the Schlicht function $F(z) = \sum\limits_{n=1}^{\infty} A_n z^n (|z| < 1)$ which is regular in the unit disc, we say that $f(z)$ is *subordinate* to $F(z)$ in the unit disc and we write $f(z) \prec F(z)$. Then there exists a bounded analytic function $\omega(z)$, regular in the unit disc, such that $\omega(0) = 0$, $|\omega(z)| \leq |z| < 1$ for which $f(z) = F(\omega(z))$.[18] The generalized Bieberbach conjecture is the assertion that

$$|a_n| \leq n|A_1| \qquad n = 1, 2, 3, \ldots \qquad (12)$$

Since each function $F(z)$ is subordinate to itself, the inequality (12) implies the Bieberbach conjecture for the class of univalent regular functions. Rogosinski,[31] has established the validity of the generalized conjecture for all $n$ in the following special cases:

    1. $F(z)$ is real on the real axis.

    2. $F(z)$ maps the unit disc onto a domain starlike with respect to the origin.

Recently, Robertson[30] has shown that the conjecture is true for all $n$, when $F(z)$ is close-to-convex in the unit disc. Further, if $F(z)$ is convex in the direction of the imaginary axis and is real on the real axis, $|a_n| \leq |A_1|$ holds for all $n \geq 1$. This extends an earlier result of Rogosinski,[31] who proves the same under the further restriction that $F(z)$ is convex in the unit disc.

Let $F(z) = \int_0^z f(t)t^{-1}dt$, where $f(z)$ is a normalized regular function univalent in $|z| < 1$. Then it was conjectured by Biernacki that $F$ was also univalent in $|z| < 1$ but Krzyz and Lewandowski disproved the conjecture.[15] Recently, Libera[16] considered the transformation

$$F^*(z) = \left(\frac{2}{z}\right)\int_0^z f(t)dt \qquad (13)$$

and proved that if $f$ is normalized, univalent, and starlike in $|z| < 1$, then so also is $F^*$. He also proved similar results when $f$ is convex and when $f$ is close-to-convex in the unit disc. Conversely, if $F(z)$ is regular and univalent in $|z| < 1$ and satisfy $F(0) = 0$ and $F'(0) = 1$, Livingston[19] has proved that the function $f(z) = 1/2[zF(z)]'$ is starlike in $|z| < 1/2$ if $F(z)$ is starlike in $|z| < 1$. If $F(z)$ is convex in $|z| < 1$, $f(z)$ is also convex for $|z| < 1/2$ and if $F(z)$ is close-to-convex for $|z| < 1$, $f(z)$ is also close-to-convex for $|z| < 1/2$. All the above results are sharp.

We now consider the following problem. Let $S^*(\alpha)$ denote the class of functions of $F(z)$ regular in the unit disc satisfying $F(0) = 0, F'(0) = 1$ and

$$\text{Re}\left\{\frac{zF'(z)}{F(z)}\right\} \geq \alpha \qquad \text{for } |z| < 1$$

where $\alpha$ is a fixed number in the interval $[0, 1]$. If $f(z) = 1/2(zf(z))'$, what is the radius $R^*(\alpha)$ of the largest disc with its center at the origin, in which $f$ satisfies the condition $\text{Re}\{zf'(z)/f(z)\} \geq \alpha$? Let $K(\alpha)$ denote the class of functions $F$ regular in $|z| < 1$, satisfying the conditions $F(0) = 0, F'(0) = 1$ and $\text{Re}\{1 + (zF''(z)/F'(z))\} \geq \alpha$, for a fixed $\alpha, 0 \leq \alpha \leq 1$, and let $f(z) = 1/2(zF)'$. What is the radius $R_c(\alpha)$ of the largest disc with the origin as center in which $f$ satisfies $\text{Re}\{1 + (zf''(z)/f'(z))\} \geq \alpha$? Recently I proved that $R^*(\alpha) = R_c(\alpha) = [\alpha - 2 + \sqrt{(\alpha^2 + 4)}]/2\alpha$ and these results are sharp.

Again let $C(\alpha, \beta)$ denote the class of normalized regular functions $F(z)$ satisfying the conditions $\text{Re}\{zf'(z)/G(z)\} > \alpha, 0 \leq \alpha < 1$; where $G(z)$ is a normalized regular function in the unit disc satisfying $\text{Re}\{zG'(z)/G(z)\} > \beta, 0 \leq \beta < 1$. If $f(z) = 1/2(zF(z))'$, what is the radius $\rho(\alpha, \beta)$ of the largest disc in which $f$ satisfies the condition $\mathrm{Re}\,\{zf'(z)/g(z)\} > \alpha$ for some $g(z)$ satisfying $\mathrm{Re}\,\{zg'(z)/g(z)\} > \beta$? The sharp estimate for $\rho(\alpha, \beta)$ turns out to be $[\beta - 2 + \sqrt{(\beta^2 + 4)}]/2\beta$, and it is interesting to note that $\rho(\alpha, \beta)$ is independent of $\alpha$.

Libera and Robertson[17] have extended the concept of close-to-convex functions to meromorphic functions

$$f(z) = \frac{1}{z} + a_0 + a_1 z + \cdots + a_n z^n + \cdots \qquad (14)$$

regular in $0 < |z| < 1$ and with a simple pole at the origin. $f(z)$, when given by (14) is called close-to-convex in the punctured disc $0 < |z| < 1$, relative to $F(z)$, if there exists a meromorphic, starlike, univalent function $F(z)$ in $|z| < 1$ with a simple pole at the origin, given by

$$F(z) = \frac{b_{-1}}{z} + b_0 + b_1 z + \cdots + b_n z^n + \cdots \ (b_{-1} \neq 0)$$

such that

$$\mathrm{Re}\,\left\{\frac{zf'(z)}{F(z)}\right\} > 0 \qquad \text{for } |z| < 1 \qquad (15)$$

It is important to note that in the meromorphic case equation (15) does not imply that $f(z)$ is univalent. For example if $f(z) = z^{-1} - 2z - 2/3z^3 + \cdots \ (0 < |z| < 1)$ and $F(z) = -1/z$, the condition (15) holds, but $f(z)$ is not univalent, since $|a_1| > 1$.

In fact, the coefficient problem for meromorphic univalent functions

$$f(z) = \frac{1}{z} + \Sigma\, a_n z^n \qquad (0 < |z| < 1) \qquad (16)$$

has attracted considerable attention. It is known that

$$|a_n| \leq \frac{2}{n+1} \qquad \text{for } n = 1, 2, \ldots \qquad (17)$$

and that $|a_3|$ may be as large as $1/2 + e^{-6}$[8] Clunie[3] has proved the validity of (17) for functions $f(z)$ starlike in $0 < |z| < 1$. In the general case, the order $0(n^{-1})$ is not the correct one for $a_n$ as Clunie[4] has pointed out. However, if $f(z)$ is univalent and close-to-convex in $0 < |z| < 1$ with a simple pole at $z = 0$, it has been established that the order $0(n^{-1})$ is correct for $a_n$.[17]

## REFERENCES

1. J. W. Alexander, "Functions which map the interior of the unit circle upon simple regions," *Ann. Math.* **17**: 12–22 (1915–1916).
2. L. Bieberbach, "Über die Koeffizienten derjenigen Potenzreihen, welche eine schlichte Abhildung des Einheitskreises vermitteln," *K. Preuss. Akad. Wiss.*, Berlin, Sitzungsbereichte, 1916, 940–955.
3. J. Clunie, "On meromorphic Schlicht functions," *J. London Math. Soc.* **34**: 215–216 (1959).
4. J. Clunie, "On Schlicht functions," *Ann. Math.* (2) 69 (1959). 511–519.
5. J. Clunie and Ch. Pommerenke, "On the coefficients of close-to-convex univalent functions," *J. London Math. Soc.*, **41**: 161–165 (1966).
6. V. F. Cowling and W. C. Royster "Some applications of the weierstrass mean value theorem," *J. Math. Soc. Japan* **13**: 104–108 (1961).
7. G. Faber, Neuer Beweis eines Koebe-Bieberbachschen Satzes über Konforme Abbildung, *K. B. Akad. Wiss. Munchen*, Sitzungesberichte der *Math. Phys. kl.*, 1916, 39–42.
8. G. K. Garabedian and M. Schiffer "A coefficient inequality for Schlicht functions," *Aun. Math.* **61**: 116–136 (1955).
9. T. H. Gronwall, "Some remarks on conformal representation," *Ann. Math.* **16**: 72–76 (1914–1915).
10. T. H. Gronwall,"Sur la deformation dans la representation conforme,"*Comtes Rendus (Paris)*, **162**: 249–252 (1916).
11. W. K. Hayman, "On successive coefficients of univalent functions," *J. London Math. Soc.* **38**: 228–243 (1963).
12. F. Herzog and G. Piranian, "On the univalence of functions whose derivatives has a positive real part," *Proc. Amer. Math. Soc.* **2**: 625–633 (1951).
13. W. Kaplan, "Close-to-convex Schlicht functions," *Michigan Math. J.* **1**: 169–185 (1952).
14. P. Koebe, "Über die Uniformiserung beliebiger analytischer Kurven," *Nach. Ges. Wiss. Gottingen*, 191–210, 1907.
15. J. Krzyz and Z. Lewandowski, "On the integral of univalent functions," *Bull. Akad. Polon. Sci. Ser. Sic. Math. Astronom. Phys.* **10**: 447–448 (1963).
16. R. J. Libera, "Some classes of regular univalent functions," *Proc. Amer. Math. Soc.*, **16**: 755–758 (1965).
17. R. J. Libera and M. S. Robertson, "Meromorphic close-to-convex functions," *Michigan Math. J.* **8**: 167–175 (1961).
18. J. E. Littlewood, "On inequalities in the theory of functions," *Proc. London Math. Soc.* **23** (2): 481–519 (1925).
19. J. E. Littlewood and R. E. A. C. Paley, "A proof that an odd Schlict functions has bounded coefficients," *J. London Math. Soc.* **7**: 167–169 (1932).
20. A. E. Livingston, "On the radius of univalence of certain analytic functions," *Proc. Amer. Math. Soc.*, **17**: 352–357 (1966).
21. K. Löwner, "Üntersuchungen uber die Verzerrung die konformen Abbildungen des Einheitskreises $|z| < 1$, die durch Funktionen mit nicht werschwindender Ableifert geleifert werden," Leipzig Ber. **69**: 89–106 (1917).

22. Z. Nehari, "Conformal Mapping," McGraw-Hill Book Co., New York, 1952.
23. R. Nevanlinna, "Über die Konforme Abbildung von Sterngebieten," *Övers. Finska Vetensk. Soc. Förth*, 63 Avd. A, No. 7 (1921-1922).
24. K. Noshiro, "On the Theory of Schlicht Functions," *J. Fac. Sci. Hokkaido Univ.*, Sapporo (I), **2**: 129-155 (1934-1935).
25. Ch. Pommerenke, "On starlike and close-to- convex functions," *Proc. London Math. Soc.* **13** (3): 290-304 (1963).
26. M. O. Reade, "Sur une classe de fonctions univalentes," *C. R. Acad. Sic. Paris* **239**: 1758-1759 (1954).
27. M. O. Reade, "On Umezawa's criteria for univalence," *J. Math. Soc. Japan* **9**: 234-238 (1957).
28. M. O. Reade, "On Umezawa's criteria for univalence II," *J. Math. Soc. Japan* **10**: 255-259 (1958).
29. M. S. Robertson, "Analytic functions starlike in one direction," *Amer. J. Math.* **58**: 465-472 (1936).
30. M. S. Robertson, "The generalized Bieberbach conjecture for subordinate functions," *Michigan Math. J.* **12**: 421-429 (1965).
31. W. Rogosinski, Über positive hamonische Entuicklungen und typischreele Potenzreihen, *Math. Z.* **35**: 93-121 (1932).
32. E. Study, "Vorlesungen über ausgewählte Gegenstande der Geometrie Zweites Heft, Konforme Abbildung einfache Zusammenhängender Bereiche," Leipzig and Berlin, 1913.
33. T. Umezawa, "On the theory of univalent functions," *Tohoku Math. J.* **7**: 212-228 (1955).
34. S. Warschawski, "On the higher derivatives at the boundary in conformal mappings," *Trans. Amer. Math. Soc.* **38**: 310-340 (1935).
35. J. Wolff, "L'integral d'une fonction holomorphe et a partie reele positive dans une demi-plan est univalent," *C. R. Acad. Sci. Paris* **198**: 1209-1210 (1934).

# Generalized Analytic Continuation

HAROLD S. SHAPIRO

*UNIVERSITY OF MICHIGAN*‡
*Ann Arbor, Michigan*

One of the most important notions in connection with the study of analytic functions is that of analytic continuation. Several recent investigations have encountered situations where there is associated with an analytic function in a domain, another function analytic in a contiguous domain which can lay claim to being a "continuation" of the original function, even though the original function is nowhere continuable in the classical sense. Such a situation arises, for instance, if the first function has nontangential limiting values almost everywhere on some smooth arc of the boundary and the second function has identical nontangential limiting values almost everywhere on that arc. As follows from a theorem of Lusin and Privalov, the two functions then uniquely determine one another. For example, if $\{z_k\}$ denotes a sequence of points which is dense on the unit circumference and $\{c_k\}$ are complex numbers such that $c_k \neq 0$, $\sum_1^\infty |c_k| < \infty$, the function

$$f(z) = \sum_1^\infty \frac{c_k}{z - z_k} \qquad |z| < 1$$

cannot be continued analytically across any point of the unit circle. Nevertheless, it has nontangential limiting values almost everywhere on $|z| = 1$, which coincide almost everywhere with the non-

---

‡Department of Mathematics.

tangential limiting values of the analytic function defined by the same series for $|z| > 1$. (A fuller discussion of this case is given in Section 2 below.)

In Ref. 3, the study of the cyclic vectors of a certain operator was found to be intimately related to the type of generalized continuation just discussed. From an altogether different point of departure, based on a study of approximation by rational functions, Gonchar (Ref. 4) was led to regard certain pairs of meromorphic functions with disjoint domains of definition as being "quasi-analytic continuations" of one another. In addition, some notion of generalized analytic continuation seems to be inherent in such diverse areas of study as the Laplace transforms of almost periodic functions, minimal solutions of interpolation problems, and overconvergence of sequences of rational functions. These considerations have prompted us to propose, in Section 1, a formal definition of generalized analytic continuation. (Note that our definition applies to *classes* of functions, rather than individual functions.) In the remaining sections, various examples are given of function classes which admit generalized analytic continuation in the sense of Section 1, or for which (as in the case of Sections 5 and 6) at least some of the features of generalized analytic continuation have been verified.

In Section 2 we discuss continuation based on "matching boundary values," already discussed briefly above. In Section 3 we shall discuss the Laplace transform of almost periodic functions and include a new proof, as well as a generalization, of an old theorem of Bochner and Bohnenblust. In Section 4 we shall recall briefly some of Gonchar's results. In Section 5 a general theorem on overconvergence, due to the author, is discussed; Section 6 has as its point of departure results of Akutowicz and Carleson on the analytic continuability of minimal solutions of interpolation problems. And, in Section 7 some open questions are listed.

The spirit of the present paper is frankly exploratory and speculative, the hope being that the juxtaposition we have made of seemingly unrelated concepts will prove fruitful. In any case, a number of new problems of a classical nature, some of them difficult, are suggested by our point of view.

The author wishes to acknowledge some stimulating conversations with his colleagues V. P. Havin (in connection with Section 2) and A. A. Gonchar (in connection with Section 4).

## 1. FORMAL DEFINITION OF GENERALIZED ANALYTIC CONTINUATION

Let $E_1, E_2$ denote disjoint open sets in the extended complex plane whose boundaries include a common Jordan arc $\alpha$. Let $M(E_i), H(E_i)$ denote, respectively, the classes of all single-valued meromorphic and holomorphic functions on $E_i$. Suppose now we have a one-one map $T$ from some subset $F_1$ of $M(E_1)$ onto a subset $F_2$ of $M(E_2)$. We shall say that $T$ defines a *generalized analytic continuation* (GAC) of the functions $f \in F_1$ into $E_2$ if the following conditions hold:

A. *Compatibility with (ordinary) analytic continuation.* If $f \in F_1$ and $f$ is analytically continuable across some point of $\alpha$, the continued function coincides with $Tf$.

B. *Permanence of functional equations.* 1) If $f$ and $cf$ belong to $F_1$, where $c$ is any complex number, then $T(cf) = cTf$. 2) If $f, g, h$ belong to $F_1$ and $f + g = h$, then $Tf + Tg = Th$. 3) If $f, g, h$ belong to $F_1$ and $fg = h$, then $(Tf)(Tg) = Th$.

*Remarks on the definition*: The properties postulated in group B seem to be (with possible exception of 3) the bare minimum we should expect of any proposed continuation. In the event $F_1$ is an algebra, $B$ may be summarized as the requirement that $T$ be an algebra homomorphism. It is clear that we could also reasonably require further properties such as the following: B4) If $f, g$ belong to $F_1$ and $g$ is the derivative of $f$, then $Tg$ is the derivative of $Tf$. It might also be reasonable to require symmetry between $F_1$ and $F_2$, that is, we could impose a requirement: C) The mapping $T^{-1}$ defines a GAC on $F_2$. If C) holds we shall say $T$ defines a *symmetric* GAC.

Actually, it is not strictly speaking essential that $E_1$ and $E_2$ should have a common boundary arc, nor even be adjacent; some results of Gonchar discussed below, suggest there are interesting situations where functions may possess a generalized continuation into a "distant" domain. In this case, we should replace A) by a requirement of compatibility with (ordinary) analytic continuation along some prescribed class of paths. We shall not, however, pursue further this line of thought.

In spite of the apparent liberality of our definition, there are *individual* functions which cannot belong to *any* class $F_1$. For example, if $f$ belongs to $H(E_1)$ and has an isolated winding point on $\alpha$, then the (ordinary) analytic continuations of $f$ across $\alpha$ on opposite sides of the winding point are already incompatible with one another, and *a fortiori* cannot both be compatible with any proposed $Tf$.

## 2. PSEUDOCONTINUATION: GAC BY MATCHING BOUNDARY VALUES

Let $E_1$ be a domain whose boundary includes a Jordan arc $\alpha$ which moreover we suppose to be *smooth*, i.e., admit a continuously turning tangent. We will say that a function $f$, meromorphic in $E_1$, *has boundary values on* $\alpha$ if at every point $z_0$ of $\alpha$, with the exception of a set of measure zero, $f(z)$ tends to a finite limit as $z$ approaches $z_0$ along every nontangential path. In this case, there is associated with $f$ a *boundary function* defined on $\alpha$ with the exception of a set of measure zero, and which, by a theorem of Lusin and Privalov (Ref. 6) uniquely determines $f$.

Suppose now that $E_2$ is a domain disjoint from $E_1$ and the boundary of $E_2$ includes $\alpha$. We shall say that a function $f \in M(E_1)$ is *pseudocontinuable across* $\alpha$ *into* $E_2$ if $f$ has boundary values on $\alpha$ and if there exists a function $g \in M(E_2)$ which has boundary values on $\alpha$ which coincide almost everywhere with those of $f \cdot g$ is then the (unique) *pseudocontinuation* of $f$ across $\alpha$. This definition was proposed by us (Ref. 7, p. 332).

Now, it is easy to verify that if we take for $F_1$ the set of $f \in M(E_1)$ which are pseudocontinuable across $\alpha$ into $E_2$, and take $T$ to be the map from $f$ to its pseudocontinuation, then $T$ defines a GAC. Indeed, property (A) follows from the Lusin–Privalov uniqueness theorem, and property (B) follows because the functional equations in question are equivalent to the corresponding functional equations for the boundary functions. Moreover, (C) holds too, so we have a *symmetric* GAC. [On the other hand, it isn't clear whether or not (B4) holds.]

At this point we encounter an embarrassing difficulty: We have not, thus far, been able to exhibit *any* meromorphic function for which we can demonstrate that it *doesn't* admit pseudocontinuation across some arc.‡ Nonetheless, we are confident that

---

‡Added in proof: We have since solved this problem; See "Functions nowhere continuable in a generalized sense," to appear in the Memorial Volume for K. Ananda-Rau, where it is shown that the function $f$ in the following theorem is not pseudocontinuable.

pseudo-continuability, when it occurs, is an exceptional phenomenon, just as is ordinary continuability. Thus, it is known that in various senses, "most" Taylor series with radius of convergence one represent nowhere continuable functions, and it seems to us plausible that one should be able to establish results of a similar nature for pseudo-continuability. One of the circumstances which makes this task seem difficult now is that very little is known about the structure of boundary functions of meromorphic functions. We have been able to exhibit a function having no pseudo-continuation *of bounded type* (in Nevanlinna's sense), namely we can show the following:

*Theorem:* Let $f(z) = \sum_{n=0}^{\infty} (z^{2^n}/2^n)$, $|z| \leq 1$. Then, there is no

function meromorphic and of bounded type (i.e., a quotient of two bounded holomorphic functions) in any subdomain of $|z| > 1$ having a sub-arc of the unit circle on its boundary and non-tangential limiting values equal to $f(e^{i\theta})$ almost everywhere on that sub-arc. (The proof of this theorem shall not be given here).

An interesting class of pseudo-continuable functions, suggested by V. P. Havin, follows. Let $E_1$ be a Jordan domain bounded by a smooth curve $\alpha$, and $m$ a complex measure concentrated on $\alpha$ which is purely singular with respect to arc length. Consider the integral $\int dm(w)/(w - z)$. It defines an analytic function inside $\alpha$, and another one outside $\alpha$. These functions are easily shown to belong to class $H^p$ in their respective domains, for each $p < 1$; hence, they possess nontangential limiting values almost everywhere on $\alpha$. Now, the singularity of $m$ can be shown to imply that the boundary values of these two functions coincide almost everywhere. Moreover, the functions are not continuable (in the ordinary sense) across any arc whose intersection with the closed carrier of $m$ is not empty. The example $\sum c_k/(z - z_k)$ discussed in the introduction is a special case of this construction. On the other hand, certain quite analogous series do not provide generalized continuations. J. Wolff (Ref. 10) showed, for instance, that it is possible to find complex numbers $\{z_k\}$, $|z_k| > 1$, $\lim |z_k| = 1$ and $\{c_k\}$, $\sum |c_k| < \infty$ such that $\sum c_k/(z - z_k)$ converges to zero for $|z| \leq 1$, and to a non-null meromorphic function for $|z| > 1$. It follows from results of Gonchar (see section 4) that such an example is not possible if $c_k$ tends to zero exponentially.

In concluding this section, we may also propose a somewhat

extended notion of continuability, which we might call *pseudo-continuability in the wide sense*, and which does not require that the function have boundary values. For simplicity, let us consider two disjoint domains $E_1$, $E_2$ which have a common frontier along an interval $\alpha$ of the real axis. Let $f$ be meromorphic in $E_1$ and $g$ meromorphic in $E_2$. If $f(z) - g(\bar{z})$ tends to zero as $z$ tends to a point $z_0$ of $\alpha$ along nontangential paths, for almost all $z_0 \in \alpha$, we can say $f$ and $g$ are pseudocontinuations of one another *in the wide sense*. Clearly, because of the Lusin–Privalov uniqueness theorem, $f$ and $g$ determine one another uniquely. One could propose other generalizations, for instance, require that $f(z) - g(2z_0 - z)$ tend to zero as $z \to z_0$ nontangentially, but this seems premature until a study of suitable examples has been made.‡

## 3. LAPLACE TRANSFORM OF ALMOST PERIODIC FUNCTIONS

In the present example $E_1$ denotes the half plane $x > 0$ ($z = x+iy$) and $E_2$ the half plane $x < 0$, $\alpha$ is the the imaginary axis. Let $G$ denote the set of Bohr almost-periodic functions, and $F_1$ that subset of $H(E_1)$ which are Laplace transforms of functions in $G$, that is

$$f(z) = \int_0^\infty \phi(t)e^{-tz}\,dt \qquad \phi \in G \qquad x > 0 \tag{1}$$

We define a map $T$ from $F_1$ into $H(E_2)$ by $Tf = g$, where

$$g(z) = -\int_{-\infty}^0 \phi(t)e^{-tz}\,dt \qquad x < 0 \tag{2}$$

Note that this map is one to one since $f \equiv 0$ implies that $\phi$ vanishes for $t > 0$ and so (by almost periodicity) for $t < 0$ so that $g \equiv 0$; and vice versa.

The compatibility requirement (A) holds, although this is nontrivial. First, we have the easily proved relation

$$\lim_{x \to 0+} xf(x + iy) = c(y) \tag{3}$$

---

‡For *holomorphic* functions it seems, on the basis of Theorem 1.11 of Zygmund's "Trigonometric Series" (vol. II, p. 204) that functions which pseudo-continue one another in the wide sense necessarily do so in the ordinary sense (we are indebted to F. Bagemihl for this remark).

(see Ref. 5), where $c(y)$ denotes the Fourier coefficient of $\phi(t)$ corresponding to the "frequency" $\lambda = y$. Therefore, if $f(z)$ is analytically continuable across some point $iy_0$ (and, therefore, bounded in a neighborhood of $iy_0$), it follows from equation (3) that for some $\epsilon > 0$, the interval $(y_0 - \epsilon, y_0 + \epsilon)$ contains no spectral exponent of $\phi$. By a theorem of Bochner and Bohnenblust (Ref. 2) (see below) $f$ and $g$ are direct analytic continuations of one another across $(y_0 - \epsilon, y_0 + \epsilon)$. This establishes (A).

As for (B), we note that (B1) and (B2) are trivially satisfied. (B3) leads to a nontrivial problem which, thus far, we have not been able to settle. In the present context, (B3) is equivalent to the proposition:

P. If $\phi_1, \phi_2, \phi_3$ are Bohr almost periodic functions which satisfy the relation

$$\phi_3(t) = \int_0^t \phi_1(u)\phi_2(t - u)\,du \qquad t > 0$$

then $\phi_1^*, \phi_2^*, \phi_3^*$ satisfy the same relation for $t > 0$, where $\phi_i^*(t) = -\phi_i(-t)$.

Indeed, $g(-z)$ for $\operatorname{Re} z > 0$ is the Laplace transform of $-\phi(-t)$, as is evident from (2). Note that in the present case the classes $F_1$ and $F_2$ stand in complete symmetry to one another.

*Remark:* We have referred to a theorem of Bochner and Bohnenblust. A new proof was given also by Peterson (Ref. 5). For completeness we shall sketch another, very short, proof based on the idea of normal families; this method is applicable in a wide variety of similar problems.

*Bochner–Bohnenblust Theorem:* If $\phi(t)$ is Bohr almost periodic and $(a, b)$ is free of points of the spectrum of $\phi$ then the functions $f(z), g(z)$ defined by equations (1) and (2) are analytic continuations of one another across the segment $(ia, ib)$.

*Proof:* Given $n > 0$, we can find a finite trigonometric sum $P_n(t) = \sum a_k e^{i\lambda_k t}$, where no $\lambda_k$ belongs to $(a, b)$ such that $|\phi(t) - P_n(t)| \leq 1/n$ for all real $t$. From this is follows at once that

$$|f(z) - R_n(z)| \leq \frac{1}{nx} \qquad x > 0 \tag{4}$$

$$|g(z) - R_n(z)| \leq \frac{1}{n|x|} \qquad x < 0 \tag{5}$$

where $R_n(z)$ is a rational function [the Laplace transform of $P_n(t)$]:

$$R_n(z) = \sum \frac{a_k}{z - i\lambda_k}$$

From (4) and (5) it follows that for all $z$, $|R_n(z)| \leq K|x|^{-1}$ where $K$ is a constant independent of $n$. Therefore, if $C$ denotes a circle with center on the interval $(ia, ib)$, the numbers

$$\int_C |R_n(z)|^{1/2} |dz|$$

remain bounded; therefore, $\{R_n\}$ which are holomorphic inside $C$ if the radius is small, form a normal family and there is a subsequence converging uniformly on compact subsets of the interior of $C$. The theorem now follows at once by comparing (4) and (5).

The considerations of this section can be extended so as to replace the Bohr almost periodic functions by the closure of the exponentials in some weighted norm. Thus, let $W(t)$ be a positive even function and define a $W$-norm for complex-valued continuous functions $\phi$ by

$$\|\phi\| = \sup \frac{|\phi(t)|}{W(t)}$$

the sup being over all real $t$.

Let $U_W$ denote the closure of the functions $\{e^{i\lambda t}\}(-\infty < \lambda < \infty)$ in the $W$-norm. We have then

*Generalized Bochner–Bohnenblust Theorem:* If $\phi \in U_W$, where

$$\int_0^\pi [\log^+ \int_0^\infty W(t)e^{-t\delta|\sin\theta|}\, dt]\, d\theta < \infty \tag{6}$$

for every $\delta > 0$, and $\phi$ is spanned (in the $W$-norm) by exponentials $\{e^{i\lambda t}\}$, where all the $\lambda$ lie outside $(a, b)$, then the Laplace transform of $\phi$ is analytically continuable across $(a, b)$ and is holomorphic in $x < 0$. The continued function is defined by equation (2).

*Remark:* It is easy to see that (6) guarantees the absolute convergence of the Laplace transform of $\phi \in U_W$.

*Proof:* We proceed just as in the previous proof, only in (4) and (5) $|x|^{-1}$ now becomes replaced by

$$\int_0^\infty W(t)e^{-t|x|}\, dt$$

Now, (6) guarantees the boundedness of the integrals

$$\int_C \log^+ |R_n(z)| |dz|$$

which ensures the normality of $\{R_n\}$ inside $C$ and the result follows.

## 4. GONCHAR'S CLASSES – HYPERCONVERGENCE

Let $E_1$ and $E_2$ be disjoint Jordan domains having a common boundary arc $\alpha$, and write $E = E_1 \cup E_2$. Let $K$ denote any set of complex numbers having no limit point in $E$. Let $\{R_n\}$ denote a sequence of rational functions, such that the degree of $R_n$ (that is, the total number of zeros and poles in the finite plane) does not exceed $n$, and $R_n$ has no poles outside $K$. Let $f$ denote a function meromorphic in the (disconnected) open set $E$. Let us suppose that $\{R_n\}$ is *hyperconvergent* to $f$, by which is meant that there exists a positive constant $p < 1$ such that for every compact subset $G$ of $E \setminus K$ there exists a constant $M = M(G)$ such that

$$\max_{z \in G} |f(z) - R_n(z)| \leq M(G) p^n \qquad n = 1, 2, \ldots$$

(Notice it is essential that $p$ be independent of the choice of $G$.) Let $f_i$ denote the restriction of $f$ to $E_i (i = 1, 2)$. It is easy to see that the $f_i$ need not be analytically continuable across $\alpha$. Gonchar (Ref. 4) showed, however, that if $f_1$ is analytically continuable across $\alpha$, the continuation must coincide with $f_2$; and vice versa. In particular, if $f_1 = 0$ then $f_2 = 0$. Of course, in view of Runge's theorem, nothing of the sort would be true without some hypothesis on the rapidity of convergence of $\{R_n\}$.

If we choose the set $K$ in all possible ways, and consider the totality of $f$ obtainable as limits of hyperconvergent sequences $\{R_n\}$, the restrictions $f_1$ of these $f$ to $E_1$ may serve as the domain of a GAC mapping $f_1$ to $f_2$, the restriction of $f$ to $E_2$. It is easy to verify, using Gonchar's results, that a symmetric GAC is thus obtained.

Not much is known about what kind of function can be the limit, even in a simply connected domain, of a hyperconvergent sequence of rational functions. This topic seems to deserve further study.

## 5. GENERALIZED CONTINUATION BY OVER-CONVERGENCE

By over-convergence we mean a phenomenon in which a sequence of functions from a certain class which converges (in a specified sense) in a given domain, necessarily converges in some other domain as well. A trivial example is afforded by functions of period one—if a sequence of these converges uniformly on compact subsets of a vertical strip of width greater than one, it converges uniformly on all compact sets.

The author has proved (see Ref. 8) a theorem which ensures over-convergence of a sequence of rational functions under certain conditions. To formulate this theorem, we consider a Banach space $S$ of functions $f(z)$ holomorphic in $E_1$: $|z| < 1$. We denote the set $|z| > 1$ by $E_2$, and impose upon $S$ the following conditions:

1. $1 \in S$, and for every $|b| > 1$, $1/(z - b)$ is an element of $S$. The set of all functions $1/(z - b)$ with $|b| > 1$ is total in $S$.

2. For every $|a| < 1$, the evaluation functional $I_a$ defined by $I_a f = f(a)$ is bounded.

3. $\lim_{b \to \infty} \|1/(z - b)\| = 0$.

4. For every $|b| > 1$, multiplication by $1/(z - b)$ is a bounded operator $M_b$ from $S$ to $S$.

5. $\|M_b\|$ is bounded on compact subsets of $E_2$.

Observe that most Banach spaces of analytic functions in $E_1$ previously studied (in particular, the $H^p$ spaces) satisfy these conditions.

*Theorem:* Suppose $S$ satisfies the above five conditions. Let $K$ be any subset of $E_2$ such that the set $U$ of functions $1/(z - b)$ with $b \in K$ is not total in $S$. Let $R$ denote the set of finite linear combinations of elements of $U$. Then, if $f_n$ is a norm convergent sequence of elements of $R$, the sequence $f_n(z)$ converges uniformly on compact subsets‡ of $E_2 \setminus \bar{K}$. Moreover, if $f_n \to 0$ in $S, f_n(z)$ tends to zero uniformly on compact subsets of $E_2 \setminus \bar{K}$.

For the proof, and extensions of this result, see Ref. 8. Roughly speaking, this theorem says that a sequence of rational functions converging "strongly" in $|z| < 1$ (that is, in the topology of the space $S$) will, if its poles are restricted to a rather "sparse" set (sparseness being determined by the hypothesis of nontotality)

---

‡By $\bar{K}$ we mean the closure of $K$.

necessarily converge also in $|z| > 1$ (over-convergence phenomenon), and so determine a meromorphic function there. This function (denote it by $\tilde{f}$) may then be thought of as a kind of continuation of $f = \lim f_n$ to $|z| > 1$. The last assertion of the theorem moreover assures that $f$ uniquely determines $\tilde{f}$.

We have not been able to determine whether or not the mapping $f \to \tilde{f}$ defines a GAC on the closure of $R$, except for an affirmative answer in some special cases (for instance, when $S = H^2$). In fact, we can't even prove the map is one to one, i.e., that $\tilde{f} = 0$ implies $f = 0$. Still less have we been able to prove compatibility with analytic continuation, in either direction. We can (under an additional hypothesis on $S$) prove the analogue of the Bochner-Bohnenblust theorem, however; if $\bar{K}$ does not contain the point $z_0$, $|z_0| = 1$, then each function $f$ in the closure of $R$ is analytically continuable across $z_0$, and the analytic continuation coincides with $\tilde{f}$ (see Ref. 8). Just as in the case of Gonchar's classes, it would be interesting to find some structural properties of functions in the closure of $R$ (always, of course, under the hypothesis of nontotality), in particular, to determine sufficient conditions for these functions to have nontangential limiting values on $|z| = 1$.

## 6. MINIMAL SOLUTIONS OF INTERPOLATION PROBLEMS

In Ref. 1, Akutowicz and Carleson proved a series of theorems of the following general type: Let $B$ denote a Banach space of analytic functions in $E_1: |z| < 1$, subject to certain further conditions (we won't enumerate here the specific results in Ref. 1, except to say the conditions on $B$ are not too restrictive, for instance, all $H^p$ spaces are permitted). Let $\{z_n\}$ be points of $E_1$, and $\{a_n\}$ complex numbers such that the interpolation problem $f(z_n) = a_n$ $(n = 1, 2, \ldots)$ has *more than one solution* in $B$. Let $K = \{1/\bar{z}_n\}$ and suppose $\beta$ is an arc of $|z| = 1$ disjoint from the closure of $K$. Then it is shown that the (unique) solution $f^*$ of the interpolation problem having minimal norm is analytically continuable across $\beta$, moreover it is indefinitely continuable in $E_2 \backslash \bar{K}$ (although not single valued, in general).

Now, the theorems of Ref. 1 are not applicable to the case when $\bar{K}$ includes the entire unit circumference. It seems to the author, by analogy with the overconvergence theorem discussed in the previous section, that one should be able to make some non-trivial assertion about $f^*$ also in the case when $\bar{K}$ includes all of $|z| = 1$. This assertion cannot now be of the nature of analytic continuability of $f^*$, but should rather be in terms of generalized continuation. We find (see Ref. 8) that when $B$ is a Hilbert space the Akutowicz–Carleson results can be reformulated in terms of overconvergence, and the theorem of the previous section becomes applicable, so that in essence the problems here are the same as those of the preceding section.

## 7. CONCLUDING REMARKS AND OPEN QUESTIONS

The preceding discussion has already pointed out some possible questions for investigation. In addition, many classical problems concerning analytic continuation may be studied for one or another type of GAC. Consider, for example, noncontinuability theorems (such as the Fabry gap theorem) from the standpoint of the type of GAC proposed in Section 2. Can one find classes of Taylor series (characterized, say, by gap conditions) which are not pseudo-continuable across any point of $|z| = 1$? That is, functions such that it is impossible to match the boundary values (a.e.) on any arc by those of a meromorphic (respectively holomorphic) function in a contiguous domain on the opposite side. Similar questions can be posed where the pseudo-continuation is required to have some further property, e.g., be of bounded type.

Another type of question is the compatibility of different types of GAC with one another—for instance, if $f$ is "continuable" across an arc both by Gonchar's method and by matching boundary values, must these continuations coincide? Perhaps some of the methods of GAC proposed in this paper are concealed instances of pseudo-continuation (either in the strict or in the wide sense). A first step towards illuminating these matters would be, as we have indicated in the preceding pages, a study of the boundary behavior of functions in the various classes under discussion [Laplace transforms of almost periodic functions, limits of hyperconvergent sequences of rational functions, limits of sequences of rational functions with sparse poles (in the sense of Section 5), solutions of minimal norm of underdetermined interpolation problems]. We wish also to emphasize the interest which, in our opinion, attaches to the problem of finding characteristics (either necessary or sufficient) for a function to be the boundary function of a meromorphic function (in the unit disc, say). The Lusin–Privalov uniqueness theorem tells us that such a function can't vanish on a set of positive measure without vanishing identically, but it would be of interest to know something about the finer structure of boundary functions. For functions of *bounded type*, there exist certain criteria due to Tumarkin, given in Chapter II, Section 8 of Ref. 6, and certain further information in Refs. 7 and 9, although even here our knowledge is very fragmentary. To the author's knowledge, nothing beyond the Lusin–Privalov uniqueness theorem, and Privalov's generalization of it (given in Ref. 6, p. 210) are known in the general case.

## REFERENCES

1. E. Akutowicz and L. Carleson, "The analytic continuation of interpolatory functions," *J. Anal. Math.* **7**: 223-247 (1959/60).
2. S. Bochner and H. F. Bohnenblust, "Analytic functions with almost periodic coefficients, *Ann. Math.* **35**: 152-161 (1934).
3. R. Douglas, H. S. Shapiro, and A. L. Shields, "On cyclic vectors of the backward shift," *Bull. Amer. Math. Soc.* **73**: 156-159 (1967).
4. A. A. Gonchar, "On quasi-analytic continuation of analytic functions through a Jordan arc," *Doklady Akad. Nauk. U. S. S. R.* **166**: 1028-1031 (1966). (Russian).
5. R. Peterson, "Laplace Transformation of Almost Periodic Functions," Eleventh Scandinavian Mathematical Congress (1949), 158-163.
6. I. I. Privalov, "Randeigenschaften Analytischer Funktionen," Deutscher Verlag der Wissenschaften, 1956, p. 212.
7. H. S. Shapiro, "Weighted polynomial approximation and boundary behaviour of analytic functions," in *Contemporary Problems of the Theory of Analytic Functions*, Nauka, Moscow (1966), p. 326-335.
8. H. S. Shapiro, "Overconvergence of sequences of rational functions with sparse poles," *Arkiv. för Matematik*, 1968 (in press).
9. H. S. Shapiro, "Smoothness of the boundary function of a holomorphic function of bounded type, and the generalized maximum principle," *Arkiv. för Matematik* 1968 (in press).
10. J. Wolff, "Sur les séries $\sum A_k/(z - \alpha_k)$," *Comptes Rendus*, **173**: 1057-1058, 1327-1328 (1921).

# Congruent Embedding into Boolean Vector Spaces

N. V. SUBRAHMANYAM

*ANDHRA UNIVERSITY*
*Waltair, India*

## 1. INTRODUCTION

Let $B = (B, +, ., ')$ be a Boolean algebra, whose *nul* and *universal* elements will be denoted by 0 and 1, respectively. By a vector space over $B$ (or simply, *a B-vector space*) is meant an additive group $V$ (whose "zero" also will be denoted by 0), together with a mapping: $(a, x) \to ax$ of $B \times V$ into $V$ such that (a) $a(x + y) = ax + ay$, (b) $(ab)x = a(bx)$, (c) $1x = x$ and (d) $(a + b)x = ax + bx$ whenever $ab = 0$. Further, a $B$-vector space $V$ is said to be *normed* if and only if to each $x \in V$ is associated a unique element $|x| \in B$ such that (a) $|x| = 0$ if and only if $x = 0$ and (b) $|ax| = a|x|$ for all $x \in V$ and $a \in B$. If $V$ is a normed $B$-vector space and we put $d(x, y) = |x - y|$, then $(V, d)$ is a $B$-metric space in the sense of the following:

*Definition:* By a *B-metric space* $S$ is meant a pair $(S, d)$, where $S$ is a set and $d: S \times S \to B$ is a mapping such that (a) $d(x, y) = 0$ if and only if $x = y$ and (b) $d(x, z) \le d(x, y) + d(z, y)$ for all $x, y, z$ in $S$.

By a *basis* of a $B$-vector space $V$ is meant a nonempty subset $G$ of $V$ such that (a) if $a_1, a_2, \ldots, a_n \in B$, $g_1, g_2, \ldots, g_n \in G$, $a_1 a_2 \ldots a_n \ne 0$ and $a_1 g_1 + a_2 g_2 + \ldots + a_n g_n = 0$, then $g_1 + g_2 + \ldots + g_n = 0$, and (b) if $x \in V$, then there exist $g_1(x), g_2(x), \ldots, g_n(x) \in G$ and $a_1(x), a_2(x), \ldots, a_n(x) \in B$ such that $a_i(x) \cdot a_j(x) = 0$ for $i \ne j$ and

$x = \sum\limits_{i=1}^{n} a_i(x)g_i(x)$. Every $B$-vector space $V$ with a basis is *necessarily* normed, although not every normed $B$-vector space has a basis (see theorem 8, Ref. 4, p. 429). If $G$ is a basis of a $B$-vector space $V$ and $x \in V$, then $|x - g| = 1$ except for a finite number of $g$'s in $G$ and $x$ has a *unique representation* as a (finite) sum $x = \sum\limits_{g \in G} a_g g$ where $a_g = |x - g|'$; also, if $g \in G$, then $|g| = 1$.

The object of this paper is to discuss some problems connected with congruent embeddings of $B$-metric spaces in a $B$-vector space with a finite basis. The results presented here resemble those of similar problems in connection with congruent embeddings of (ordinary) metric spaces in the Euclidean spaces. (See Ref. 1; part II, Chapter IV.) Further, they extend the same type of solutions of Melter[3] for a $p$-ring to the case of any $B$-vector space with a finite basis. The proofs given here are much shorter than those of Melter even for the case of a $p$-ring.

One of the tools which we employ below is the concept of "inner product." If $V$ is any normed $B$-vector space (with or without a basis) and $x, y \in V$, then $|x - y| \leq |x| + |y|$ so that there is a unique solution $(x, y)$ of the two equations $\eta|x - y| = 0$ and $\eta + |x - y| = |x| + |y|$. Also, if $V$ has a finite basis $g_1, g_2, \ldots, g_k$ and we write $x = \sum\limits_{i=1}^{k} a_i g_i$ and $y = \sum\limits_{i=1}^{k} b_i g_i$ where $x, y \in V$ and $a_i a_j = b_i b_j = 0$ for $i \neq j$, then we have $(x, y) = \sum\limits_{i=1}^{k} a_i b_i$. The reader is referred to Refs. 4 and 5 for details of these and other concepts used in the presentation below.

## 2. AN ISOMETRY THEOREM

Let $V$ be a vector space with a finite basis $G$ over a Boolean algebra $B$; and let $k$ denote the number of elements in $G$. First we recall[5] the following:

*Lemma 1:* If $y_1, y_2, \ldots, y_n \in V$, $n < k$ and $a > \sum\limits_{i=1}^{n} |y_i|$, then there exists an element $z \in V$ such that $|z| = a$ and $|y_i - z| = a$ for $1 \leq i \leq n$.

*Corollary 1:* If $y_0, y_1, \ldots, y_n \in V$ and $n < k$, then there exists an element $y \in V$ such that $|y - y_i| = 1$ for $0 \leq i \leq n$.

*Lemma 2:* Any equilateral set contains at most $k + 1$ elements.

*Corollary 2:* A maximal equilateral set has exactly $k + 1$ elements.

Since any basis of $V$ together with $0 \in V$ is a maximal equilateral set (of *side* 1), it follows that every basis of $V$ has precisely $k$ elements, so that we may call $k$ the *dimension* of $V$. In fact, Jagannadham[2] has shown that even in the infinite case, any two bases of $V$ must have the same cardinal number. Further, if $x_0, x_1, x_2, \ldots, x_k$ is any (maximal) equilateral set of side 1 and $x \in V$, then, by lemma 2, $\prod_{i=0}^{k} |x - x_i| = 0$. A maximal equilateral set $E$ of side 1 is called an *equibasis* of $V$ since each $x \in V$ can be uniquely written in the *normal* form $x = \sum_{e \in E} a_e e$ where $a_e a_f = 0$ for $e \neq f$ and $\sum_{e \in E} a_e = 1$; in fact, we must have $a'_e = |x - e|$ (See Ref. 5; lemma 5, p. 405). Now we shall prove the following:

*Theorem 1:* A $B$-metric space $S$ is isometric to a finite dimensional $B$-vector space if and only if

1.1: $S$ contains a finite equilateral set $E$ of side 1, where $E$ has at least two elements,

1.2: $S$ can be congruently embedded in every $B$-vector space in which $E$ can be so embedded, and

1.3: If $p, q \in S$ and $a \in B$, then there exists $r \in S$ such that $d(r, p) = a\, d(p, q)$ and $d(r, q) = a'\, d(p, q)$.

We need some lemmas to prove this theorem.

*Lemma 3:* Let $p_0, p_1, \ldots, p_n$ $(n > 0)$ be an equilateral set of side 1 in a $B$-metric space $S$, and assume 1.3 of theorem 1 above. Further, let $a_0, a_1, \ldots, a_n$ be a partition of $B$, that is, $a_0, a_1, \ldots, a_n \in B$, $a_i a_j = 0$ for $i \neq j$ and $a_0 + a_1 + \ldots + a_n = 1$. Then there exists an element $p \in S$ such that $d(p, p_i) = a'_i$ for $0 \leq i \leq n$.

*Proof:* Obvious for $n = 1$; and now assume the result for $n-1$. Since $a_0, a_1, \ldots, a_{n-2}, a_{n-1} + a_n$ also is a partition of $B$, there exists $q \in S$ such that $d(q, p_i) = a'_i$ for $0 \leq i \leq n - 2$ and $d(q, p_{n-1}) = a'_{n-1}$ $\cdot a'_n$. Now, if $i \neq n$, $1 = d(p_n, p_i) \leq d(p_n, q) + d(q, p_i)$ and, hence,

$$1 = \prod_{i=0}^{n-1} [d(p_n, q) + d(q, p_i)] = d(p_n, q) + \prod_{i=0}^{n-1} d(q, p_i) = d(p_n, q)$$

Hence, there exists $p \in S$ such that $d(p, q) = a_n$ and $d(p_n, p) = a'_n$. Now since $a_i\, d(q, p_i) = 0$ for $0 \leq i \leq n - 1$ and $a_i a_n = 0$ for $i \neq n$, it follows by triangle inequality that $a_i\, d(p, p_i) = 0$ for all $i$; also, $d(p, p_i) + d(p, p_j) = 1$ for $i \neq j$. Hence the lemma. Q. e. d.

*Lemma 4:* Let $x_0, x_1, \ldots, x_n$ be an equilateral set of side 1 in a normed $B$-vector space (with or without a basis) and let $\{a_i | 0 \leq i \leq n\}$ and $\{b_i | 0 \leq i \leq n\}$ be partitions of $B$. If $x = \sum_{i=0}^{n} a_i x_i$ and $y = \sum_{i=0}^{n} b_i x_i$, then $|x - y| = \sum_{i=0}^{n} (a_i b_i' + a_i' b_i)$.

We omit the simple proof of this lemma. (See Ref. 5; corollary 7, p. 406.)

*Proof of theorem 1:* Assume 1.1, 1.2, and 1.3, and let $E = \{p_0, \ldots, p_n\}$. Then $E$ is congruent to any equibasis of a $B$-vector space $V$ of dimension $n$ and hence, by 1.2, $S$ can be congruently embedded in $V$. Let $\varphi : S \to V$ be any congruent embedding. If $p \in S$, then $\prod_{i=0}^{n} d(p, p_i) = \prod_{i=0}^{n} |\varphi(p) - \varphi(p_i)| = 0$, since $\varphi(E)$ is a maximal equilateral set of side 1 in $V$. Also, $d(p, p_i) + d(p, p_j) = 1$ for $i \neq j$ and $d'(p, p_i)\varphi(p) = d'(p, p_i)\varphi(p_i)$ for all $i$. Hence,

$$\varphi(p) = \sum_{i=0}^{n} d'(p, p_i)\varphi(p) = \sum_{i=0}^{n} d'(p, p_i)\varphi(p_i)$$

Now, if $x \in V$, then there is a partition $\{a_i | 0 \leq i \leq n\}$ of $B$ such that $x = \sum_{i=0}^{n} a_i \varphi(p_i)$ and, by lemma 3 above, there is an element $p \in S$ such that $d(p, p_i) = a_i'$ for $0 \leq i \leq n$. Since we now have $\varphi(p) = x$, it follows that $\varphi$ is an isometry.

To prove the converse, it is sufficient to show that every $B$-vector space $V$ of (finite) dimension $n$ satisfies 1.1, 1.2, and 1.3. Now 1.1 is obvious and 1.2 follows by lemma 4. Finally, let $x, y \in V$ and $a \in B$; if $z = ax + a'y$, then $|x - z| = a'|x - y|$ and $|y - z| = a|x - y|$. q. e. d.

*Corollary 3:* Any two $B$-vector spaces of the same (finite) dimension are isometric to each other.

As a $B$-metric space of the most general type which is congruent to a finite dimensional $B$-vector space, we mention the following example borrowed from an idea of Foster:

*Example:* Let $S$ be the set of all $(n + 1)$-tuples $(n > 0)$

$$a = (a_1, a_2, \ldots, a_{n+1})$$

where $a_i a_j = 0$ for $i \neq j$ and $\sum_{i=1}^{n+1} a_i = 1$; and define

$$d(a, b) = \sum_{i=1}^{n+1} (a_i b_i' + a_i' b_i)$$

Then $S$ is congruent to a $B$-vector space of dimension $n$.

## 3. EMBEDDING THEOREMS

Let $V$ be a $B$-vector space of dimension $k$; then the corollary 1 above implies that any *metric basis* of $V$ (that is, any subset $G$ of $V$ such that $|x - g| = |y - g|$ for all $g \in G$ implies $x = y$) must contain at least $k$ elements. We now prove

*Lemma 5:* A $k$-tuple of $V$ is a metric basis of $V$ if and only if it is an equilateral set of side 1.

*Proof:* Let $g_1, g_2, \ldots, g_k$ be a metric basis of $V$. By corollary 1 and the fact that any translate of a metric basis also is a metric basis, we may assume that $|g_i| = 1$ for $1 \le i \le k$. Now let $e_1, e_2, \ldots, e_k$ be a basis of $V$ and write, for $1 \le i \le k$,

$$g_i = \sum_{j=1}^{k} a_{ij} e_j \quad \text{where} \quad a_{ij} a_{ik} = 0 \quad \text{for} \quad j \ne k$$

Also, for each $k$-tuple $(j) = (j_1, j_2, \ldots, j_k)$ of integers between 1 and $k$, put $c_{(j)} = a_{1j_1} a_{2j_2}, \ldots, a_{kj_k}$ and observe that $c_{(j)} g_i = c_{(j)} e_{j_i}$. Suppose now that $j_1, j_2, \ldots, j_k$ are *not all distinct* and let $e_{(j)}$ denote one of the basis elements different from all the $e_{j_i}$, $i = 1, 2, \ldots, k$; and put $z = c_{(j)} e_{(j)}$. Then

$$|z - g_i| = |c_{(j)}\{e_{(j)} - g_i\} - c'_{(j)} g_i| = |c_{(j)}\{e_{(j)} - e_{j_i}\} - c'_{(j)} g_i|$$
$$= c_{(j)}|e_{(j)} - e_{j_i}| + c'_{(j)}|g_i| = c_{(j)} + c'_{(j)} = 1 = |g_i - 0|$$

for $1 \le i \le k$, so that $z = 0$ since $g_1, g_2, \ldots, g_k$ is a metric basis; so $c_{(j)} = |z| = 0$. Hence,

$$1 = \prod_{i=1}^{k} |g_i| = \prod_{i=1}^{k} \{\sum_{i=1}^{k} a_{ij}\} = \sum c_\sigma$$

where the summation is over all permutations $\sigma$ of integers $1, 2, \ldots, k$. Now assume $m \ne n$. Then for each permutation $\sigma$ we have

$$c_\sigma |g_m - g_n| = |c_\sigma g_m - c_\sigma g_n| = c_\sigma |e_{\sigma(m)} - e_{\sigma(n)}| = c_\sigma$$

so that, summing both sides over all permutations $\sigma$, we obtain $|g_m - g_n| = 1$.

The proof of the converse can be found in Ref. 5. Q. e. d.

If $g_1, g_2, \ldots, g_k$ is a metric basis of $V$, it now follows that there is a unique $g_0$ such that $g_0, g_1, \ldots, g_k$ is an equibasis of $V$. Consequently, if $g_1, g_2, \ldots, g_k$ and $h_1, h_2, \ldots, h_k$ are any two metric bases of $V$, there is a unique isometry $T: V \to V$ such that $Tg_i = h_i$ for $1 \le i \le k$. (See Ref. 5; p. 409–410.) Now we shall prove the following:

*Theorem 2:* Let $S$ be a $B$-metric space containing a $k$-tuple which is congruent to a metric basis of $V$. If every $(k + 2)$-tuple of $S$ can be congruently embedded in $V$, then $S$ itself can be so embedded in $V$.

*Proof:* We use the standard symbol $\approx$ to denote congruence, and, further, the elements on the left of this symbol below are from $S$, while those on the right will be in $V$.

Now let $p_1, p_2, \ldots, p_k$ be a $k$-tuple of $S$ congruent to a metric basis of $V$. This then is an equilateral set of side 1 in $S$.

Now suppose that $p_1, p_2, \ldots, p_k \approx g_1, g_2, \ldots, g_k$, and let $q \in S$. Also, suppose that $p_1, p_2, \ldots, p_k, q \approx h_1, h_2, \ldots, h_k, y$. Then there is a unique isometry $T: V \to V$ such that $Th_i = g_i$ for $1 \leq i \leq k$. Put $\lambda(q) = Ty$. Suppose now that we have also $p_1, p_2, \ldots, p_k, q \approx h'_1, h'_1, \ldots, h'_k, y'$ and that $T'$ is the unique isometry such that $T'h'_i = g_i$ for $1 \leq i \leq k$. Then

$$|T'y' - g_i| = |y' - h'_i| = d(q, p_i) = |y - h_i| = |Ty - g_i|$$

for $1 \leq i \leq k$, and since $g_1, g_2, \ldots, g_k$ is an equilateral set of side 1, it is a metric basis. Hence $Ty = T'y'$ so that $\lambda: S \to V$ is well defined.

Finally, suppose that $q, r \in S$, and let

$$p_1, p_2, \ldots, p_k, q, r \approx h_1, h_2, \ldots, h_k, y, z$$

If now $T$ is the isometry that carries $h_i$ into $g_i$ for $1 \leq i \leq k$, then $\lambda(q) = Ty$ and $\lambda(r) = Tz$; hence, $|\lambda(q) - \lambda(r)| = |y - z| = d(q, r)$. Q. e. d.

In connection with the hypothesis in the above theorem, it is interesting to note the following:

*Theorem 3:* Any $B$-metric $n$-tuple, where $n \leq k + 1$, can be congruently embedded in $V$.

We first prove a simple lemma:

*Lemma 6:* If $a, b, c$ are the sides of a triangle in a $B$-metric space and $e \in B$, then $eab + ea'b' + c = e + c$ and $ab'c' = 0$.

*Proof:* Observe that $a + b = a(b + c) + b(c + a) = ab + bc + ca = ab + c(a + b) = ab + c$ so that $eab + ea'b' + c = e(ab + c + a'b') + e'c = e(a + b + a'b') + e'c = e + e'c = e + c$, and the other part is trivial. Q. e. d.

*Proof of theorem 3:* The construction of the embedding is due to Melter, but the proof given here is much shorter and uses the concept of inner product in $V$ (See Ref. 3; p. 1003–1005).

Let $p_0, p_1, \ldots, p_n$ be elements of a $B$-metric space, where $n \leq k$, and put $a_{ij} = d(p_i, p_j)$. Further, let $g_1, g_2, \ldots, g_k$ be a basis of $V$ and write

$$x_0 = 0$$

$$x_1 = a_{01}g_1$$

$$x_2 = a_{02}a'_{12}g_1 + a_{02}a_{12}g_2$$

$$x_3 = a_{03}a'_{13}g_1 + a_{03}a_{13}a'_{23}g_2 + a_{03}a_{13}a_{23}g_3$$

$$\cdots\cdots\cdots\cdots\cdots\cdots\cdots\cdots\cdots\cdots\cdots$$

$$x_j = a_{0j}a'_{1j}g_1 + a_{0j}a_{1j}a'_{2j}g_2 + \ldots + a_{0j}a_{1j}\ldots a_{j-2,j}a'_{j-1,j}g_{j-1}$$
$$+ a_{0j}a_{1j}\ldots a_{j-1,j}g_j$$

$$\cdots\cdots\cdots\cdots\cdots\cdots\cdots\cdots\cdots\cdots\cdots$$

$$\cdots\cdots\cdots\cdots\cdots\cdots\cdots\cdots\cdots\cdots\cdots$$

Obviously, $|x_i - x_0| = |x_i| = a_{0i} = d(p_0, p_i)$ for $1 \leq i \leq n$; and now let $0 < i < j \leq n$. Then

$$(x_i, x_j) = a_{0i}a_{0j}a'_{1i}a'_{1j} + \ldots + a_{0i}a_{0j}a_{1i}a_{1j}\ldots a'_{i-1,i}a'_{i-1,j}$$
$$+ a_{0i}a_{0j}a_{1i}a_{1j}\ldots a_{i-1,i}a'_{ij}$$

so that, by repeated application of lemma 6, we have $a_{ij}(x_i, x_j) = 0$ and $a_{ij} + (x_i, x_j) = a_{0i}a_{0j} + a_{ij} = a_{0i} + a_{0j} = |x_i| + |x_j|$; hence, $|x_i - x_j| = a_{ij} = d(p_i, p_j)$. Q. e. d.

Finally, we shall prove the following theorem, which again is connected with theorem 2:

*Theorem 4:* An arbitrary $(k + 2)$-tuple $(p_0, p_1, \ldots, p_{k+1})$ of a $B$-metric space can be congruently embedded in $V$ if and only if $\prod_{i \neq j} d(p_i, p_j) = 0$.

We require two lemmas to prove this theorem.

*Lemma 7:* Let $g \in V$, and assume $|g| = 1$. Further, let $U$ be the set of all $x \in V$ such that $(g, x) = 0$. Then $U$ is congruent to a $B$-vector space of dimension $k - 1$.

*Proof:* Since $|g| = 1$, there is an equibasis $g_0, g_1, \ldots, g_k$ of $V$ with $g_0 = 0$ and $g_k = g$. And it follows that $U$ is the set of all $x \in V$ which can be written in the form $x = \sum_{i=1}^{k-1} a_i g_i$ where $a_i a_j = 0$ for $i \neq j$. It is now easy to complete the proof by using theorem 1. Q. e. d.

*Lemma 8:* If $x_0, x_1, \ldots, x_k \in V$ and $\prod_{i \neq j} |x_i - x_j| = 0$, then there exists an element $g \in V$ such that $|g - x_i| = 1$ for $0 \leq i \leq k$.

*Proof:* Let $J$ be the set of integers $0, 1, 2, \ldots, k$ and let $P$ be the set of all mappings of $J$ into $J$. Further, let $e_0, e_1, \ldots, e_k$ be an equibasis of $V$ and write, for $0 \le i \le k$, $x_i = \sum_{j=0}^{k} a_{ij}e_j$, where $a_{ij}a_{im} = 0$ for $j \ne m$ and $\sum_{j=0}^{k} a_{ij} = 1$. Also, for each $\sigma \in P$, put $c_\sigma = a_{0\sigma 0} \cdot a_{1\sigma 1}a_{2\sigma 2} \ldots a_{k\sigma k}$ and observe that $\{c_\sigma | \sigma \in P\}$ is a partition of $B$. (See the proof of lemma 1 Ref. 5, p. 402.) Further, if $\sigma \in P$ is a permutation, and $m \ne n$, then $c_\sigma(x_m - x_n) = c_\sigma[e_{\sigma m} - e_{\sigma n}]$ and, hence, $c_\sigma < |x_m - x_n|$; and hence $c_\sigma = 0$. Now for each nonpermutation $\sigma \in P$, let $e_\sigma$ denote one of the $e$'s different from $e_{\sigma i}, i = 0, 1, 2, \ldots, k$, and put $g = \sum c_\sigma e_\sigma$. Then this element $g$ has the desired properties. Q. e. d.

Now theorem 4 follows easily from the lemmas 7 and 8.


## REFERENCES

1. L. M. Blumenthal, "Theory and Applications of Distance Geometry," Clarendon Press, Oxford, 1953.
2. P. V. Jagannadham, "Linear transformations in a boolean vector space," *Math. Ann.*, **167**: 240–247 (1966).
3. R. A. Melter, "Contributions to Boolean Geometry of $p$-Rings, *Pacific. J. Math.* **14**: 995–1017 (1964).
4. N. V. Subrahmanyam, "Boolean vector spaces I," *Math. Z.* **83**: 422–433 (1964).
5. N. V. Subrahmanyam, "Boolean vector spaces II," *Math. Z.* **87**: 401–419 (1965).

# Raikov Systems

J. H. WILLIAMSON

*UNIVERSITY OF CAMBRIDGE*
*Cambridge, England*

---

This topic has its origin in a study by D. A. Raikov, about 1940, of the structure of the measure algebra $\mathfrak{M}(R)$ of the real line $R$. Perhaps the most convenient account of this work is to be found in Sections 31–33 of Ref. 2. It is still true that the main interest of the subject is due to its bearing on the structure of $\mathfrak{M}(R)$ and, more generally, of $\mathfrak{M}(G)$, where $G$ is a locally compact abelian topological group; this structure is far from being fully understood, even today. However, there is also a case for studying Raikov systems for their own sake, independent of any possible applications to $\mathfrak{M}(R)$. They provide yet another instance, at a quite elementary level, of just how complicated a system the real numbers are.

The ideas and results described here—insofar as they are not already well known—represent work still in progress, and certainly do not constitute a finished theory. It is hoped, however, that enough can be said to justify the claim that the subject is of substantial interest and that it would amply repay further study. In order to simplify matters, only the case of the real line is considered; as in very many situations where the group structure of $R$ plays a part, an extension to a general locally compact abelian group $G$ is in most respects straightforward.

To begin with we recall some standard notation and terminology. If $A_1, \ldots, A_n$ are subsets of $R$, we write $A_1 + \cdots + A_n$ for their vector sum $\{a_1 + \cdots + a_n : a_1 \in A_1, \ldots, a_n \in A_n\}$. If $A_1 = \cdots$ $= A_n = A$, we write $(n)A$ for the vector sum; it is convenient to introduce the convention $(0)A = \{0\}$ for any nonempty subset $A$ of $R$. The set $\{a - t : a \in A\}$, where $t$ is a fixed element of $R$, is denoted $A - t$; more generally $A_1 - A_2$ is the set $\{a_1 - a_2 : a_1 \in A_1, a_2 \in A_2\}$, and $-A$ is $\{-a : a \in A\}$. The set-theoretic difference of $A_1$ and $A_2$ is denoted $A_1 \backslash A_2$.

A subset of $R$ is of type $F_\sigma$ if it is a countable union of compact sets. If $A$ is of type $F_\sigma$ then so are $-A$ and $A - t$ ($t \in R$); if $A_1$ and $A_2$ are of type $F_\sigma$ so are $A_1 + A_2$ and $A_1 - A_2$, and if $A_1, \ldots, A_n$ are all of type $F_\sigma$, so is $A_1 + \cdots + A_n$. All of these results are immediate consequences of the continuity of the algebraic operations in $R$. A set is of the first category in $R$ if it is a countable union of nowhere dense subsets of $R$. A set that is both of type $F_\sigma$ and of Lebesgue measure zero is necessarily of the first category in $R$. This follows from the fact that a compact (hence, closed) set of measure zero must be nowhere dense. A subset of $R$ will be called perfect if it is compact and has no isolated points.

We can now say what we mean by a Raikov system of subsets of $R$. These are called in Ref. 2 and elsewhere "regular systems," but the term "regular" is so much overworked that it seems desirable to replace it by something more specific. A nonempty collection $\mathscr{F}$ of subsets of $R$, of type $F_\sigma$, will be called a Raikov system if it satisfies the following conditions:

R1. If $A_1 \in \mathscr{F}$, and $A_2$ is a subset of $A_1$ of type $F_\sigma$, then $A_2 \in \mathscr{F}$.

R2. If the countable collection of sets $A_1, A_2, \ldots$ are all in $\mathscr{F}$, then so is their union.

R3. If $A \in \mathscr{F}$ and $t \in R$, then $A - t \in \mathscr{F}$.

R4. If $A \in \mathscr{F}$, then $(2)A \in \mathscr{F}$.

Evidently (in view of R1 and R2), we could replace R1 to R4 by R1, R2, R3, and

R4'. If $A_1 \in \mathscr{F}$, $A_2 \in \mathscr{F}$, then $A_1 + A_2 \in \mathscr{F}$.

It follows at once that a Raikov system contains the vector sum of any finite collection of sets of the system.

If the system also satisfies the following:

R5. If $A \in \mathscr{F}$, then $-A \in \mathscr{F}$.

we shall call it a symmetric Raikov system. Not all Raikov systems are symmetric. (See Ref. 2, Section 32)

It is clear that every Raikov system contains all countable sets; indeed the class of all countable sets is the simplest example of a Raikov system. It is symmetric. Another example—also symmetric—is the class of all subsets of $R$ of type $F_\sigma$. This we shall call the improper system; any other system is proper. By maximal we shall mean maximal among the proper Raikov systems of the kind under consideration (e.g., symmetric systems).

*Proposition 1:* The following are equivalent:

(i) $\mathscr{F}$ is proper;

(ii) $R \notin \mathscr{F}$;

(iii) $m(A) = 0$ for all $A \in \mathscr{F}$, where $m$ is Lebesgue measure on $R$.

Although the fact of the existence of maximal Raikov systems is not required in the applications to $\mathfrak{M}(R)$, such systems and their properties are of interest for their own sake and raise one or two problems.

*Proposition 2:* Given any proper (proper symmetric) Raikov system, there exists a maximal (maximal symmetric) Raikov system containing it.

*Proof:* This is a straightforward application of Zorn's maximal principle.

One open problem in this connection is whether every maximal Raikov system is necessarily symmetric. It seems likely that the answer is negative, and that there exist maximal systems that are contained in no proper symmetric system. There is an interesting note by Erdös (Ref. 1) in which he produces a semigroup $S$ in $R$ that is of measure zero, but which generates $R$ ($S - S = R$). The proof involves the assumption of the continuum hypothesis as well as of the axiom of choice. The semigroup $S$ would seem to be very far from being an $F_\sigma$ set, and there may well be no proper Raikov system containing it. However, it leads one to suppose that there may exist semigroups, with more agreeable properties, that are of measure zero but are not contained in any proper subgroup of $R$. Another possible approach to the problem is as follows. One can find perfect sets $A$ such that $m(A - A) > 0$, and $m[(n)A] = 0$ for $n = 2$ (quite easily) and for $n = 3$ (with more difficulty). It is reasonable to conjecture that such sets can be constructed for all positive integers $n$. If this were so then it would imply the existence of nonsymmetric maximal Raikov systems. On the other hand, such systems might conceivably exist even though sets with the property indicated did not.

The intersection of any collection of Raikov systems is again a Raikov system. It follows that, given any collection $\{A_i\}$ of $F_\sigma$ sets, there is a minimal Raikov system (and a minimal symmetric Raikov system) containing them; we have only to take the intersection of all systems (all symmetric systems) containing the sets $A_i$. If $\mathscr{F}$ is the minimal Raikov system containing $\{A_i\}$, then we say that $\{A_i\}$ is a set of generators of $\mathscr{F}$. Thus, for example, the system of all countable sets is generated by any single-point set, and the improper system by any set of type $F_\sigma$ and strictly positive Lebesgue measure.

*Proposition 3:* If $\mathscr{F}$ has a countable set of generators then it has a single generator.

Raikov systems with a single generator are in many respects easier to deal with than general systems. It will appear presently (Proposition 8, Corollary) that there exist systems that do not have a single generator. It is frequently useful to know that if there is a single generator then it can be taken to be a subset of $R$ of some special type. More generally, we have the following:

*Proposition 4:* In any Raikov system, each generator may be assumed to be either

(i) a compact set, or

(ii) a semigroup, or, in the case of a symmetric Raikov system,

(iii) a group.

*Proof:* Suppose that $A = \cup_{r=1}^\infty A_r$ is a generator of $\mathscr{F}$, where each $A_r$ is compact. Each $A_r$ may be assumed to be a subset of an interval $[a_r, b_r]$, where $a_r \in A_r$. Dissect $[a_r, b_r]$ into a number $n_r$ of sub-intervals $[a_r^0, a_r^1], \ldots, [a_r^{n_r-1}, a_r^{n_r}]$ (where $a_r^0 = a_r, a_r^{n_r} = b_r$), so that the length of each of them is not greater than $r^{-1}$. Write

$$A_r^s = A_r \cap [a_r^{s-1}, a_r^s]$$

then

$$A' = \cup_{r=1}^\infty \cup_{s=1}^{n_r} (A_r^s - a_r^{s-1})$$

is a compact set that can be used as a replacement for $A$. Clearly, any Raikov system that contains one must contain the other. To see that $A'$ is compact, it is only necessary to note that if $\{V_i\}$ is any open covering of $A'$, then $0 \in V_{i_0}$ for some $i_0$, and $A' \setminus V_{i_0}$

is a finite union of compact sets (since we have chosen our dissections so as to become smaller and smaller as $r$ increases).

If $A$ is a generator, then $A' = \cup_{r=1}^{\infty} (r)A$ is also a generator, and is a semigroup. If the Raikov system is symmetric then $A'' = \{0\} \cup \cup_{r=1}^{\infty} (r)(A \cup -A)$ is a group that can be used in place of $A$.

If $\{A_i\}$ is a set of generators of $\mathscr{F}$, the sets of $\mathscr{F}$ can be described in terms of the sets $\{A_i\}$ as follows: A set is in $\mathscr{F}$ if and only if it is an $F_\sigma$ subset of a set of the form

$$\cup_{n=1}^{\infty} [(r_{k_1}) A_{i_{k_1}} + \cdots + (r_{k_n}) A_{i_{k_n}} - t_{k_1,\ldots,k_n}]$$

This may simplify considerably in special cases. Thus, if there is a single generator $A$, the sets of $\mathscr{F}$ are the $F_\sigma$ subsets of sets of the form

$$\cup_{n=1}^{\infty} [(r_n) A - t_n];$$

if the single generator $A$ is a semigroup then the sets of $\mathscr{F}$ are the $F_\sigma$ subsets of sets

$$\cup_{n=1}^{\infty} (A - t_n)$$

that is, of countable unions of translates of $A$.

We now introduce a notion of generalized independence for subsets of $R$. Let $E$ be a given subset of $R$; we shall say that the set $X$ is independent with respect to $E$, or $E$-independent, if the relation

$$\sum_{r=1}^{N} n_r x_r \in E \qquad (n_r \in Z, x_r \in X)$$

is possible only if $0 \in E$ and $n_r = 0$ for $1 \leq r \leq N$. This reduces to the usual definition of independence for subsets of $R$ when $E = \{0\}$.

*Proposition 5:* Let $E$ be a given set of the first category and let $I$ be a given interval; then there exists a perfect $E$-independent subset of $I$.

*Proof:* We imitate the standard construction (Ref. 3, p. 20–21). Suppose that $E = E_1 \cup E_2 \cup \ldots$, where each $E_r$ is nowhere dense; we may assume without loss of generality that $E_1 \subset E_2 \subset \cdots$.

At the first stage, choose two closed subintervals $B_1^{(1)}$ and $B_2^{(1)}$ of $I$ such that $B_1^{(1)} \times B_2^{(1)}$ does not cut any line

$$n_1 x_1 + n_2 x_2 = a_1$$

with $a_1 \in E_1$ and $|n_1| \leq 1, |n_2| \leq 1$. This is always possible, since

$E_1$ is nowhere dense in $R$. At the $j$-th stage, if intervals $B_1^{(j-1)}, \ldots,$ $B_{2^{j-1}}^{(j-1)}$ are initially present, choose $B_1^{(j)}$ and $B_2^{(j)}$ to be closed sub-intervals of $B_1^{(j-1)}, \ldots, B_{2^j-1}^{(j)}$ and $B_{2^j}^{(j)}$ closed sub-intervals of $B_{2^{j-1}}^{(j-1)}$, so that $B_1^{(j)} \times \cdots \times B_{2^j}^{(j)}$ does not cut any hyperplane

$$n_1 x_1 + \cdots + n_{2^j} x_{2^j} = a_j$$

where $a_j \in E_j$ and $|n_r| \leq j$ for $1 \leq r \leq 2^j$; and also so that the length of $B_r^{(j)}$ does not exceed $j^{-1}$ for $1 \leq r \leq 2^j$. This can always be achieved, since $E_j$ is nowhere dense.

Write $A^{(j)} = \cup_{r=1}^{2^j} B_r^{(j)}$, and $A = \cap_{j=1}^{\infty} A^{(j)}$. Evidently $A$ is a perfect subset of $I$. Given $N$ distinct points of $A$, since the lengths of the intervals $B_r^{(j)}$ tend to zero it follows that if $j$ is large enough the points will be in $N$ distinct sets of the form $B_r^{(j)}$. Hence, no linear relation of the form

$$\sum_{r=1}^{N} n_r x_r = a$$

can hold if $a \in E_j$ and $|n_r| \leq j (1 \leq r \leq N)$. But this implies that no linear relation of this kind can hold for any choice of $a \in E$ and $n_r \in Z$, which is what was required.

We now explore one or two consequences of the above proposition.

*Proposition 6:* Let $H$ be a subgroup of $R$ of type $F_\sigma$ and of measure zero, and $P$ a perfect $H$-independent subset of $R$. Then the group generated by $H$ and $P$ is also of type $F_\sigma$ and of measure zero.

*Proof:* Write $Q = P \cup (-P)$; then the group required is

$$H + \cup_{r=0}^{\infty} (r) Q$$

this is easily seen to be of type $F_\sigma$. It may equally well be written as

$$\cup_{r=0}^{\infty} [H + (r) Q]$$

each set $H + (r)Q$, being of type $F_\sigma$, is measurable, and in order to prove that the group is of measure zero it is enough to prove that $m[H + (r)Q] = 0$ for each $r$.

Suppose that $m[H + (k)Q] > 0$; then $H + (2k)Q$ contains an open neighborhood, say $V$, of $0$ in $R$ (as is shown by the argument of Ref. 2, p. 181, taking into account the fact that everything is symmetric about zero). Let $S$ be the set of points $(p_1, \ldots, p_{2k+2})$ in $R^{2k+2}$ with $p_r \in P (1 \leq r \leq 2k + 2)$ such that $h + \sum \epsilon_r p_r \in V$ for some choice of $h \in H$ and of $\epsilon_r = \pm 1 (1 \leq r \leq 2k + 2)$. For each set $\{\epsilon_r\}$ and each $h$ the map

$$(p_1, \ldots, p_{2k+2}) \to h + \sum \epsilon_r p_r$$

is continuous, and so $S$ is an open subset of $P^{2k+2}$. It is not empty since there exist $h$ and $\epsilon_1, \ldots, \epsilon_{2k}, +1, -1$ such that

$$h + \sum_{r=1}^{2k} \epsilon_r p_r + p - p \in V$$

Now, since $V \subset H + (2k)Q$, we have, if $(p_1, \ldots, p_{2k+2}) \in S$,

$$\sum_{r=1}^{2k+2} \pm p_r + h_1 = h_2 + \sum_{r=1}^{2k} \pm p'_r$$

But now if the $p_r$ are all different we have a nontrivial linear relation of the form $\sum \pm p_r \in H$, which is not possible since $P$ is $H$-independent. It follows that $S$ is contained in a finite union of sets of the form $\{(x_1, \ldots, x_{2k+2}) : x_i \pm x_j = 0\}$. But no open set in $P^{2k+2}$ can be contained in such a union, since $P$ is perfect. For every neighborhood of each point of $P$ contains infinitely many distinct points of $P$, and we could then find, in an arbitrary neighborhood of any point of $S$ a point of $S$ with $p_1, \ldots, p_{2k+2}$ all different; this is clearly a contradiction.

We can deduce one or two consequences from Proposition 6:

*Corollary 1:* Let $P$ be closed and without isolated points; the conclusion of Proposition 6 still holds.

*Proof:* We can write $P = \cup_{r=1}^{\infty} P_r$, where each $P_r$ is perfect. Since each $P_r$ is $H$-independent, the group generated by $H$ and $P_r$ is of measure zero for each $r$, but the group generated by $H$ and $P$ is just the union of all these, and so is of measure zero also. It is evidently of type $F_\sigma$.

*Corollary 2:* Let $P$ be a closed independent subset of $R$, with $P = P_1 \cup P_2$, where $P_1$ is countable, and $P_2$ is closed and has no isolated points; then the group generated by $P$ is of measure zero.

*Proof:* Since $P_1$ is countable, so is the group $H$ generated by $P_1$. It is clear that $P_2$ is $H$-independent; as in Corollary 1 the group generated by $H$ and $P_2$ is of measure zero. But this is just the group generated by $P$.

The next result is also a consequence of Proposition 6, but is somewhat more substantial, so we state it as a separate proposition:

*Proposition 7:* Let $P_1$ be a given perfect independent set; then there exists another such set $P_2$ such that $P_1 \cap P_2 = \phi$ and $P_1 \cup P_2$ is again perfect and independent.

*Proof:* If $P_1$ is perfect and independent, the group $H$ generated by $P_1$ is well known to be of measure zero (see, for instance, Ref. 4,

p. 108; the result is also a special case of Proposition 6). The group $H$ is then of the first category, since it is of type $F_\sigma$. Take $H$ to be the set $E$ of Proposition 5, and let $P_2$ be the set $P$ constructed so as to be $H$-independent. Then clearly $P_1 \cup P_2$ is perfect, since $P_1$ and $P_2$ are. It is also independent, for if

$$\sum n_r x_r + \sum n'_r x'_r = 0$$

$(n_r, n'_r \in Z, x_r \in P_1, x'_r \in P_2)$ we would have $\sum n'_r x'_r \in H$, hence $n'_r = 0$, all $r$, hence $\sum n_r x_r = 0$, hence $n_r = 0$, all $r$, since $P_1$ is independent.

*Proposition 8:* Given any proper symmetric Raikov system with a single generator, there exists a strictly larger proper symmetric Raikov system.

*Proof:* If the system $\mathscr{F}$ is symmetric and has a single generator, we may suppose that this generator is a group, say $H$ (Proposition 4). Since $\mathscr{F}$ is proper, $m(H) = 0$ and so $H$ is of the first category. By Proposition 5 there exists a perfect $H$-independent set $P$, and by Proposition 6 the group $H_1$ generated by $H$ and $P$ is of measure zero. But any set in the Raikov system $\mathscr{F}_1$ generated by $H$ and $P$ is a subset of a countable union of translates of $H_1$, and so is of measure zero. Thus, by Proposition 1, the system $\mathscr{F}_1$ is proper; it evidently properly contains $\mathscr{F}$.

*Corollary:* A maximal symmetric Raikov system must have uncountably many generators.

There are various problems that arise in connection with the above results; perhaps the most obvious are to examine the possibility of removing the restrictions "symmetric" and "with one generator" in certain places. Another is to examine the possibility of generalizing by considering a pair of Raikov systems, one strictly contained in the other; one might then look for sets of the second system that are independent with respect to a generating group or semigroup of the first system.

We turn now to the question of the relation of Raikov systems to measures on $R$. By measure we mean regular complex Borel measure with finite total mass. Writing $\|\mu\| = \int |d\mu|$, these measures form a Banach space $\mathfrak{M}(R)$ which can be identified with the dual of the Banach space $\mathfrak{C}_0(R)$ (the complex continuous functions vanishing at infinity) with the uniform norm. $\mathfrak{M}(R)$ can be made into a Banach algebra by writing

$$(\mu * \nu)(E) = \int \mu(E - t)\, d\nu(t)$$

for each Borel set $E$; or alternatively, if the measures are regarded primarily as functionals, by writing

$$(\mu * \nu)(f) = \iint f(s + t) \, d\mu(s) \, d\nu(t)$$

for $f \in \mathfrak{C}_0(R)$.

Now, given a Raikov system $\mathscr{F}$, we say that $\mu$ is concentrated on $\mathscr{F}$ if for any Borel set $E$ we have

$$|\mu|(E) = \sup |\mu|(A) \qquad \text{for } A \subset E, A \in \mathscr{F}$$

Similarly, if the right side of this last relation is zero, we say that $\mu$ is concentrated outside $\mathscr{F}$. Notice that to say that $\mu$ is concentrated on $\mathscr{F}$ is not to say that its support is a set of $\mathscr{F}$; take, for example, a measure with mass $r^{-2}$ at the $r$-th rational. The support of this is the whole real line, but it is concentrated on $\mathscr{F}_0$, the Raikov system of countable sets.

If $\mu$ is a non-negative real measure, write

$$\mu_1(E) = \sup \mu(A) \qquad \text{for } A \subset E, A \in \mathscr{F}$$

It is then clear that $\mu_1$ is a measure concentrated on $\mathscr{F}$, and $\mu_2 = \mu - \mu_1$ is concentrated outside $\mathscr{F}$. Hence, by linearity we have a unique decomposition for any measure:

$$\mu = \mu_1 + \mu_2$$

where $\mu_1$ is concentrated on $\mathscr{F}$ and $\mu_2$ is concentrated outside $\mathscr{F}$; we have also

$$\|\mu\| = \|\mu_1\| + \|\mu_2\|$$

We may regard $\mu_1$ as the "projection" of $\mu$ on $\mathscr{F}$. Now it is a rather remarkable fact, but not hard to prove (Ref. 2, Section 31) that the measures of $\mathfrak{M}(R)$ that are concentrated on $\mathscr{F}$ form a closed sub-algebra $\mathfrak{M}(\mathscr{F})$ of $\mathfrak{M}(R)$, and the measures concentrated outside $\mathscr{F}$ form a closed ideal $\mathfrak{M}(\mathscr{C}\mathscr{F})$ in $\mathfrak{M}(R)$. Not every decomposition of $\mathfrak{M}(R)$ into a closed sub-algebra and a complementary ideal is of this form; see Ref. 2, Section 33.

A well-known classical theorem of Wiener and Pitt (Ref. 5) asserts—*inter alia*—that when $\mathfrak{M}(\mathscr{F}_0)$ is the algebra of atomic measures in $\mathfrak{M}(R)$, there exists a measure $\nu$ that is strongly independent of $\mathfrak{M}(\mathscr{F}_0)$ in the sense that the measures $\mu_r * \nu^r$ and $\mu_s * \nu^s$ are mutually singular if $r \neq s$, whatever measures $\mu_r, \mu_s \in \mathfrak{M}(\mathscr{F}_0)$ are chosen. This has the consequence that

$$\|\mu_0 + \mu_1 * \nu + \cdots + \mu_n * \nu^n\| = \|\mu_0\| + \|\mu_1 * \nu\| + \cdots + \|\mu_n * \nu^n\|$$

and, if one takes a little care in the construction of $\nu$, it is possible to assert the existence of a measure $\nu$ such that for any measure $\mu \in \mathfrak{M}(\mathscr{F}_0)$ with $\|\mu\| \leq \|\nu\|$, $\mu + \nu$ has no inverse. In fact, one can make a stronger assertion, but this one will suffice for our present purpose. For details, see Ref. 5 or 6. We may say that the Wiener–Pitt phenomenon occurs between $\mathfrak{M}(\mathscr{F}_0)$ and $\mathfrak{M}(R)$. On the basis of the results described above, we have the following:

*Proposition 9:* If $\mathscr{F}$ is a proper symmetric Raikov system with a single generator, then the Wiener–Pitt phenomenon occurs between $\mathfrak{M}(\mathscr{F})$ and $\mathfrak{M}(R)$.

The proof of this result is much as in the classical case (see, e.g., Ref. 6) but is rather more complicated in certain respects, and the details will be published elsewhere.

It is very likely that further improvements in the classical results could be made, in various directions. One obvious condition to try to drop is "symmetric"; it is quite likely that this can be done. It seems less plausible that the condition "with a single generator" can be dispensed with easily. Another direction in which progress might be made is to replace $\mathfrak{M}(R)$ by $\mathfrak{M}(\mathscr{F})$ for some suitable Raikov system $\mathscr{F}$. It would then be possible to assert the following:

*Proposition 9′:* If $\mathscr{F}_1$ and $\mathscr{F}_2$ are proper symmetric Raikov systems, each with a single generator, and such that $\mathscr{F}_1$ is properly contained in $\mathscr{F}_2$, then the Wiener–Pitt phenomenon occurs between $\mathfrak{M}(\mathscr{F}_1)$ and $\mathfrak{M}(\mathscr{F}_2)$.

This is as yet unproved, but seems likely to be true.

One way of looking at the Wiener–Pitt phenomenon is to regard it as an undesirable, or pathological, feature of the algebra $\mathfrak{M}(R)$. One would wish to be able to identify "large" sub-algebras of $\mathfrak{M}(R)$ in which the phenomenon does not occur. The present investigations do not bear directly on this problem; they show rather that the pathology is likely to be spread uniformly throughout the algebra $\mathfrak{M}(R)$ (at least, if Proposition 9′ holds). More refined results would show, presumably, that the pathology is more uniform; it would be of considerable interest to know, for example, whether $\mathfrak{M}(\mathscr{F}_1)$ and $\mathfrak{M}(\mathscr{F}_2)$ could be replaced by any two closed sub-algebras of $\mathfrak{M}(R)$, similarly related to each other. But in order to settle this question it would appear that methods quite different from those described here would be required.

## REFERENCES

1. P. Erdös, "On some properties of Hamel bases," *Colloq. Math.* **10**: 267–269 (1963).
2. I. M. Gelfand, D. A. Raikov, and G. E. Shilov, "Commutative Normed Rings," Chelsea Publishing Co., New York, 1964.
3. J.-P. Kahane and R. Salem, "Ensembles parfaits et séries trigonométriques," Hermann, Paris, (1963).
4. W. Rudin, "Fourier Analysis on Groups," Interscience Publishers, New York, 1962.
5. N. Wiener and H. R. Pitt, "Absolutely convergent Fourier-Stieltjes transforms," *Duke Math. J.* **4**: 420–436 (1938).
6. J. H. Williamson, "Banach algebra elements with independent powers and theorems of Wiener-Pitt type," in "Functional Algebra," Scott, Foresman and Co., Chicago, 1966, pp. 186–197.

# Certain Problems in the Design of Programmed-Motion Systems

## A. S. GALIULLIN‡

*PEOPLE'S FRIENDSHIP UNIVERSITY*
*Moscow, U.S.S.R*

---

The fundamental notions and axioms of classical mechanics, as well as the methods developed therein, have been directed primarily toward the solution of the following problems:[1,2]

1. Given the effective forces acting on a mechanical system and the constraints imposed on the particles of the system and on the system as a whole, determine the law of motion governing the system (the direct problem).

2. Given the motion of a mechanical system, determine the forces acting on that system (the converse problem).

It is reasonable that these problems should have constituted the basic objectives of theoretical mechanics for so long, inasmuch as these are the problems to which man's practical activity has required satisfactory solutions. However, even in the course of the mathematical formulation of these problems and their diverse modifications, another objective of theoretical mechanics has made itself prominent, namely the problem of constructing mechanical systems whose motions have predetermined properties, e.g., the properties of stability or optimality in some sense.[3-7,19-23]

Now this problem has long since emerged beyond the sphere of mere theoretical mechanics; it has become transformed into

---

‡Department of Theoretial Mechanics.

the major problem of process control, i.e., the problem of designing such systems invested with various physical qualities and
structures, wherein processes take place in compliance with previously imposed requirements.[7-12]

One of the main mathematical objectives in this problem area
is the formulation of differential equations to fit a prescribed
particular solution, to fit prescribed particular integrals, or in
general to meet prescribed attributes of a certain particular solution.

In the present article we consider some possible statements
of problems in the theory of programmed-motion system design,
with particular reference to systems, mechanical, electrical, or
otherwise, whose motions follow a predetermined program.

The staff of the Department of Theoretical Mechanics of the
Patrice Lumumba Friendship of Nations University (UDN im. P.
Lumumby) and the members of the perpetual scientific seminar
associated with the Department are currently directing their efforts
toward the solution of problems of this type (see Refs. 13, 16,
19, and 23 and the master's theses of I. A. Mukhametzyanov,
A. G. Aleksandrov, Zh. Kirgizbaev, and R. G. Mukharlyamov).

We are interested in systems whose motions are described by
ordinary differential equations

$$\dot{y} = f(y, u, \dot{u}, t) \tag{1}$$

where $y[y_1, \ldots, y_n]$ is the state vector of the system (generalized
coordinates of the system), and $u[u_1, \ldots, u_m]$ is the control vector
of the system (internal parameters and controlling forces).

The motion program is assumed given, either in the form of
a programmed state vector for the system

$$\varphi[\varphi_1(t), \ldots, \varphi_n(t)] \tag{2}$$

comprising the laws of variation governing the coordinates of the
system in programmed motion (elements of the program), or in
the form of a certain integral manifold

$$\omega(y, t) = 0 \tag{3}$$

consisting of the particular integrals

$$\omega_x(y, t) = 0 \qquad (x = 1, \ldots, k; k \leq n) \tag{4}$$

The actual problem of designing programmed-motion systems
may be formulated mathematically in the following forms:[13]

## A. PROGRAMMING OF THE TIME VARIATION OF THE SYSTEM PARAMETERS

Given a system of differential equations

$$\dot{y} = f(y, u, \dot{u}, t) \qquad (5)$$

determine the control vector $u(t)$ such that this system will have the prescribed particular solution

$$y = \varphi(t) \qquad (6)$$

or the prescribed integral manifold

$$\omega(y, t) = 0 \qquad (7)$$

This statement embraces problems of determining the laws of variation of the control parameters of physical systems whose structure is known in the large.

Stated in this form, for example, is a problem in the dynamics of a variable-mass point,[5] namely, given the equation of motion of a variable-mass point $m(t)$ in a certain given force field $\bar{F}$:

$$\frac{d}{dt}(m\bar{v}) = \bar{F} + \bar{u}\frac{dm}{dt} \qquad (8)$$

where $\bar{v}$ and $\bar{u}$ are the absolute velocities of the point itself and of the escaping mass, determine the mass of the point and the velocity of the escaping mass such that the motion of the point will follow a prescribed law

$$\bar{r} = \bar{r}(t) \qquad (9)$$

The following problem in rocket dynamics may also be stated in the same form:[14]

Given the equations of motion of a heavy nonrotating rocket in a vertical plane, based on the usual tenets of exterior ballistics,

$$m\frac{d\bar{v}}{dt} = \bar{T} + \bar{R} + m\bar{g}$$

$$\frac{d}{dt}(I\omega) = M_a + M_{rd} + M_c \qquad (10)$$

where $m\bar{g}$ is the weight of the rocket, $\bar{v}$ is the velocity of the center of gravity of the rocket, $\bar{T}$ is the tractive force, $\bar{R}$ is the principal aerodynamic force vector acting in the plane of motion of the rocket, $I\omega$ is the angular momentum of the rocket about the transverse axis passing through the center of gravity of the rocket perpendicularly to the plane of motion, $M_a$, $M_{rd}$, and $M_c$ are the aerodynamic moment, reaction damping moment, and control moment about the same transverse axis, respectively, determine the time variation of the steering moment such that the motion of the center of gravity of the rocket will follow a trajectory prescribed in the vertical plane.

## B. PROGRAMMING OF THE CLOSURE OF A SYSTEM

Given a system of differential equations

$$\dot{y} = f(y, u, t) \tag{11}$$

construct a system of closing equations

$$\dot{u} = g(y, u, t) \tag{12}$$

such that the closed system obtained as a result will admit the prescribed particular solution

$$y = \varphi(t) \tag{13}$$

or the prescribed integral manifold

$$\omega(y, t) = 0 \tag{14}$$

This statement embraces problems of determining the equations for the controls and feedback loops in physical systems executing programmed motion. The structure of the controlled plant in this case and all of its internal parameters are assumed to be known. Problems of this type in the design of programmed-motion systems are referred to in the literature as problems in the analytical design of controllers.[15]

The following problem in rocket dynamics may be stated, for example, in this form:[16]

Given the equations of motion of a controlled rocket in a vertical plane:

$$\ddot{y} = F_1(\dot{y}, z, \dot{z}, \beta, \delta, m, \dot{m}, c)$$
$$\ddot{z} = F_2(\dot{y}, z, \dot{z}, \beta, \delta, m, \dot{m}, c) \tag{15}$$
$$\ddot{\beta} = F_3(\dot{y}, z, \dot{z}, \dot{\beta}, \beta, \delta, m, \dot{m}, c)$$

where $y$ and $z$ are the range and height of the center of gravity of the rocket, $m$ is the mass of the rocket, $c$ is the effective jet velocity, $\beta$ is the angle of inclination of the longitudinal axis of the rocket with respect to the horizontal axis, and $\delta$ is the angle of deflection of the control surfaces, determine the equation

$$\ddot{\delta} = F(\delta, \dot{\delta}, \beta, \dot{\beta}, t) \tag{16}$$

for the controls such that the motion of the center of mass of the rocket is ensured along a trajectory specified in the vertical plane.

## C. DESIGN OF THE TOTAL PROGRAMMED-MOTION SYSTEM (GENERAL PROBLEM)

The problem is to construct a system of differential equations

$$y = f(y, t) \tag{17}$$

according to a given particular solution

$$y = \varphi(t) \tag{18}$$

or according to a given integral manifold

$$\omega(y, t) = 0 \tag{19}$$

We note that whereas the structure of the physical system itself was assumed given in the preceding problems (the equations of motion of the actual system were given), in this problem the structure of the system is not specified beforehand, and the problems is reduced to one of finding precisely this structure of the system in the form of differential equations describing the motion of the system itself and the controls.

Consequently, the given problem of programmed-motion system design is a logical generalization of the preceding problems, and it reduces to the familiar problem of constructing a set of systems of differential equations having a prescribed particular solution.[17]

Stated in this form, for example, is the following problem in the programmed control of metalworking machine tools:[18]

Given a surface

$$F(x, y, z) = 0 \tag{20}$$

of a piece to be machined, determine the analytical structure

$$
\begin{aligned}
\dot{x} &= f_1(x, y, z) \\
\dot{y} &= f_2(x, y, z) \\
\dot{z} &= f_3(x, y, z)
\end{aligned}
\tag{21}
$$

of the differential analyzer controling the motion of the machine tool along a particular curve on this surface.

All of these programmed-motion system design problems have one feature in common in all their multivarious forms, and that is the fact that if they do have a solution,‡ the solution is not unique.

In problems of programming by variation of the parameters of the system itself, this nonuniqueness is principally a result of the fact that the control elements $u_\mu$ are determined from the necessary conditions for realizability of the motion of the system according to the given program, which are obtained by replacing the $y_\nu$ in the equations of motion by the corresponding program elements $\phi_\nu(t)$, and these conditions are normally differential equations in the control elements $u_\mu$. Of course, the number of these equations may prove smaller than the number of control elements.

The multivalued character of the solution of the programmed-motion system design problem is boldly outlined in the general statement of the problem. For example, if the program is specified in the form of an integral manifold

$$\omega_i(y_1, \ldots, y_n) = 0 \qquad (i = 1, \ldots, s; \, s < n) \tag{22}$$

the corresponding unknown system of equations has the form[19]

$$\dot{y}_j = \frac{1}{\Delta}\left[\sum_{i=1}^{s} \Delta_{ij}\phi_i(\omega, y) - \sum_{k=s+1}^{n} \Delta^{jk} P_k(y)\right] \qquad (j = 1, \ldots, s)$$

$$\tag{23}$$

$$y_m = P_m(y) \qquad\qquad\qquad\qquad (m = s + 1, \ldots, n)$$

where $\Delta_{ij}$ is the cofactor of the $(i, j)$th element in the determinant

$$\Delta = \frac{\partial(\omega_1, \ldots, \omega_s)}{\partial(y_1, \ldots, y_s)}$$

and $\Delta^{jk}$ is the determinant derived from $\Delta$ when the $j$-th column $(j = 1, \ldots, s)$ is replaced by the $k$th column $(k = s + 1, \ldots, n)$ of the matrix $\partial(\omega_1, \ldots, \omega_s)/\partial(y_1, \ldots, y_n)$.

We note that the only conditions imposed on the functions $\phi_i(\omega, y)$ and $P_k(y)$ in the system of equations (23) constructed above are those relating to the existence and uniqueness of solution, apart

---

‡It is possible that a solution may not exist, if for no other reason, because the motion prescribed for the given physical system is not realizable in general. For example, a variable-mass point under gravitational pull cannot execute motion along a straight line when the velocities of the point itself and the escaping mass are collinear.[5]

from the necessary requirement that $\phi_i(0, y) = 0$ $(i = 1, \ldots, s)$, but otherwise all of these functions are entirely arbitrary and cannot be predetermined within the scope of the stated problem.

Consequently, for the final description of the control parameters, the controls, and the actual programmed-motion system as a whole it is necessary to impose additional conditions, both on the elements of the system and on the motion program.

These auxiliary conditions may take the form of initial values of the control elements, restrictions on the values of the control elements on some time interval, optimality requirements on the programmed motion in one sense or another, some kind of general engineering demands on the system, etc.

Let us suppose that all the elements of the programmed-motion system have been defined in one way or another, for example, by the application of the auxiliary conditions indicated above.

With this supposition, however, the motion of the system according to a specified program is only possible in the event the initial values of the system coordinates exactly coincide with the initial values of the program and as long as there are no steady perturbing forces present or preturbations of the system parameters (parametric perturbations).

In actual fact, of course, both initial and steady perturbations, as well as parameteric perturbations, are always present.

Therefore, in the design of programmed-motion systems it is important to bear in mind another group of additional requirements, namely the requirement of program stability in the face of initial, steady, and parametric perturbations.

We point out that the programmed motion of a system in the general case is specified conditionally as a certain set of indeterminate functions $\varphi_v(t)$ or $\omega_x(y, t)$. Then the optimality and stability conditions, which complete the description of the solution of any of the given problems, will also be imposed on the program elements. These conditions must be taken into account when designating the specific program of motion for the physical system.

## REFERENCES

1. J. L. Lagrange, "Analytical Mechanics" [Russian translation], Gostekhizdat, 1950.
2. G. K. Suslov, "Theoretical Mechanics" [in Russian], Gostekhizdat, 1946.

3. N. E. Zhukovskii, "Stability of Motion, Collected Works, Vol. I" [in Russian], Gostekhizdat, 1948.

4. A. M. Lyapunov, "General Problem of the Stability of Motion" [in Russian], Gostekhizdat, 1950.

5. I. V. Meshcherskii, "Papers on the Mechanics of Variable-Mass Bodies" [in Russian], Gostekhizdat, 1949.

6. L. S. Polak, "Variational Principles of Mechanics" [in Russian], Gostekhizdat, 1960.

7. K. Lantsosh, "Variational Principles of Mechanics" [in Russian], Izd. "Mir," 1965.

8. H. S. Tsien, "Engineering Cybernetics," McGraw-Hill Book Co., N. Y., 1954.

9. A. G. Ivakhnenko, "Engineering Cybernetics" [in Russian], Gostekhizdat, UkrSSR, 1959.

10. A. M. Letov, "Stability of Nonlinear Control Systems" [in Russian], Fizmatgiz, 1962.

11. A. A. Fel'dbaum, "Fundamentals of the Theory of Optimal Automatic Systems" [in Russian], Izd. "Nauka," 1966.

12. R. E. Bellman and S. E. Dreyfus, "Applied Dynamic Programming," Princeton University Press, Princeton, N. J., 1962.

13. A.S. Galiullin, "Dynamic programming problems," *Tr. UDN im. P. Lumumby, seriya teor. mekh.* **5** (2): (1964).

14. L. Davis, Jr., W. Follin, Jr., and L. Blitzer, "Exterior Ballistics of Rockets," Van Nostrand, Princeton, N. J., 1958.

15. A. M. Letov, "Analytical controller design," *Avtomatika i telemekhanika* **21** (4): (1960).

16. A. S. Galiullin, "Analytical design of rocket control systems," *Tr. UDN im. P. Lumumby, seriya teor. mekh.* **1** (1): (1963).

17. N. P. Erugin, "Construction of a complete set of systems of differential systems having a specified integral curve," *Prikl. matem. i mekh.*, **16** (6): (1952).

18. M. B. Ignat'ev, "Holonomic Automatic Systems," [in Russian], (Izd. AN SSSR, 1963).

19. R. G. Mukharlymov, "Construction of a Set of Systems of Differential Equations Having Specified Integrals," Differentsial'nye uravneniya, No. (1966).

20. I. A. Mukhametzyanov, "Construction of a set of systems of differential equations of stable motion according to a specified program, *Tr. UDN sm. P. Lumumby. seriya teor. mekh.* **1** (1): (1963).

21. A. G. Aleksandrov, "Optimality of automatic control systems in the problem of automatic controller design," *Tr. UDN im. P. Lumumby, seriya teor. mekh.* **17** (4): (1966).

22. Zh. Kirgizbaev, "Stability of the programmed motion of a variable-mass point, taking account of the earth's rotation," *Izv. Akad. Nauk Kaz. SSR, Ser. Fiz. Mat.* No. 1 (1966).

23. A. S. Galiullin, "Certain Problems in the Stability of Programmed Motion," [in Russian], Tatknigoizdat, 1960.

# Author Index

# Subject Index